STATICS OF STRUCTURAL COMPONENTS

STATICS OF STRUCTURAL COMPONENTS

UNDERSTANDING BASIC STRUCTURAL DESIGN

GIUSEPPE DE CAMPOLI, Ph.D., P.E.
City College of the
City University of New York

A Wiley-Interscience Publication

JOHN WILEY & SONS

New York Chichester Brisbane Toronto Singapore

Library of Congress Cataloging in Publication Data:

de Campoli, Giuseppe.
Statics of structural components.

"A Wiley-Interscience publication."
Includes index.
1. Structural design. 2. Statics. I. Title.
TA658.D4 1983 624.1'771 82-20122
ISBN 0-471-87169-9

Printed in the United States of America

10 9 8 7 6 5 4 3 2 1

To Dean Richard M. Bossone,
in appreciation of his guidance

PREFACE

Erecting the structure is the final stage of the creative process that begins with the architect's intuition of the most functional and pleasing building for the activities intended therein. However, we are not born with intuitive ability; it results, rather, from a combination of love for the task at hand and systematic practice. And loving the task is a necessary but insufficient condition for achievement: Only rigorous and constant training harnesses creative forces into productive ability. The greatest artists, writers, composers, conductors, or musicians could not give the world the enjoyment of their art without training. Pablo Picasso, in nearly a century-long life of artistic activity, continued to train in the painstaking traditional ways of drawing and painting before beginning those works that were to establish him as one of the geniuses of our age.

It is therefore deceptive to think that structural design is an exception to the rule and can be taught by mere intuition. If free artistic expression of human creativity requires systematic practice to surface into beautiful forms or sounds, then technical creativity, or understanding it, cannot be born of effortless superficiality.

Granted, simple visual demonstrations of physical principles using familiar objects can and should be provided as an aid in recalling theories. Visual aids alone, however, cannot be a substitute for teaching general principles, since students may fail to link specific cases with general rules. It would be easy for the teacher (and this would be popular among students) to reduce instruction in structural design to a discussion of the behavior of a few simplified models. But with such an approach, it is very difficult, if not impossible, for teachers and students alike to obtain a true and general understanding of structural behavior, let alone intuitive building creativity.

For one, without analytic expressions, the focus of a teacher's discussion is easily lost. A formula, like a picture, says more than a thousand words; both should be used in teaching structures. There is another risk in casual or informal teaching: Like a bright but short-lived flare, students' interest may be alerted and then lost forever. Nothing valuable and lasting is achieved with little involvement and effort.

Great architectural geniuses have conceived and built daring structures before structural theories and computational aids existed. It would suffice to recall Michelangelo's dome on St. Peter's Cathedral in Rome. Michelangelo, however, was a genius. Moreover, with the commission, he was given limitless materials, labor, and time. Most probably, he also made several tests on scale models. Today, architects, engineers, and builders must perform similarly impressive tasks with a minimum of construction materials and the least amount of labor possible in the shortest amount of time in order to reduce the cost of the structure and produce early returns for the investor who commissioned the structure. This is why even a great, intuitive genius of structural behavior like Pier Luigi Nervi would never build structures without a precise numerical analysis of their stress pattern and values, nor without, on occasion, testing his intuition on scale models.

Once intuitively guided structural designers begin creating and building structures in the contemporary world, then the possibility, let alone the superiority, of intuitive teaching can be substantiated. Until then, books such as this will offer an understanding of the basics of structural design with a straightforward approach, and an abundance of visual aids and numerical examples.

GIUSEPPE DE CAMPOLI

New York, New York
February 1983

CONTENTS

STATICS OF STRUCTURAL COMPONENTS

The model of an elevated reservoir, an award-winning structure designed by the author. Photograph by G. Gherardi-A. Fiorelli, Rome.

ONE

THE STEPS OF A SYSTEMATIC APPROACH

Designing structures usually proceeds in the following sequence. We discuss each step in detail later.

1. Selecting the structural scheme or model structure.
2. Evaluating boundary forces or reactions of the structure's constraints.
3. Evaluating inner forces.
4. Evaluating cross-sectional properties of the structure's members.
5. Evaluating inner stresses.
6. Comparing inner stresses with the safe stresses specified by building codes.
7. Evaluating deflections and rotations at relevant points of the structure.
8. Comparing actual deformations with the limit values specified by building codes.

Large structures are usually designed with the aid of the computer, in which case the preceding steps would constitute a flow chart for the computer program used.

The aim of this book is to give architecture and engineering students an early and clear understanding of structural behavior and structural design techniques. Automated analysis is therefore not discussed here. It is left, instead, to later courses that emphasize the skills required for the routine application of concepts learned at this stage. The title *Statics of Structural Components* indicates to what extent structural design is discussed.

Structural components considered in this book are statically determinate beams, cables, trusses, and arches. Therefore, Step 2 does not include evaluating statically indeterminate reactions. Step 3 is applied to the previously mentioned one-dimensional and plane structural components, excluding such two-dimensional structures as plates loaded in their own mid-plane or at right angle to it, as well as three-dimensional structures; that is, grids, space frames, cable nets, membranes, and shells. Step 4 concludes the discussion of statics, since evaluating cross-sectional properties involves applying the same operational techniques of force vectors used for statics (Steps 2 and 3) to area scalars. Evaluating internal stresses (Step 5) and structural deformations (Step 7) is left to textbooks on strength of materials, while the discussion of two- and three-dimensional structures will be found in textbooks on advanced structural design.

(Overleaf)
9 West 57th Street, New York City. The effect of the trusses shown in the narrow elevation of the building is explained in Figure 2.27 and the related discussion. Architects: Skidmore, Owings, and Merrill–New York Office. Structural Engineers: Paul Weidlinger and Associates, New York. Photograph by Wolfgang Hoyt/ESTO.

TWO

THE MODEL STRUCTURE

It is obvious that a schematic of the real structure must be drawn from the onset of structural design. The real structure is initially only an idea, and even in the case of an existing structure, it would be impossible to work with its real size, real loads, and real support conditions. The model structure, or structural scheme, is described by three sets of data: geometries, constraints, and loads.

2.1. GEOMETRIES OF THE MODEL STRUCTURE

Geometric data must include the spans of floors and roofs, the rise of arches and domes, the sag of cables and membranes, the height of columns, the coordinates of bar ends for trusses and space frames, and all pertinent information necessary to completely describe the position of all points of a structure in space.

Architectural structures are usually made of framed elements whose length far exceeds the dimensions of their cross section, such as beams, columns, arches, cables, bars of trusses, and space frames. In the model structure, therefore, these elements are simply represented by the curved or straight lines to which the centers of all cross sections of the element belong. Such lines are called the geometric axes of the structural element, and they are drawn in a conveniently reduced scale. If the structure is made of plates, shells, or membranes, the thickness is the negligible dimension, and the coordinates of the points of a grid lying on the midsurface of the structure are used in the model.

2.2. CONSTRAINTS OF THE MODEL STRUCTURE

The importance of completely understanding constraint conditions, or connections, between elements of a structure and also between the total structure and its foundation cannot be overemphasized. Correctly modeling constraints gives greater reliability to the results of structural analysis. Moreover, being aware of the influence that various constraint conditions have on the static behavior of buildings enables the architect-engineer to condition that behavior efficiently and aesthetically.

Before substantiating these statements with examples, we must clarify the meaning of structural constraint. The elements of a structure perform their load-carrying function according to a hierarchy of importance that is easy to identify. Flat or ribbed floor and roof panels discharge their loads on beams, which are, therefore, the constraints of those panels: They hold the panels in place. Beams are often supported by larger, more important beams called girders; they are the constraints of the beams. Girders are, in turn, supported by vertical structures, such as walls, piers, or columns, which are the constraints of the girders. Vertical structures are supported on foundation footings, pile caps, caissons, grade beams, and general mats. These foundations are supported by the ground with or without the help of foundation piles.

The true constraints that each structural element exerts on those next in the hierarchy are only approximated in the structural scheme, since their behavior is often complex. It is important, however, to use the best possible approximation. It is clear, moreover, from our discussion, that most constraints are man-made; even foundation strata can be improved by drainage and consolidation. The architect-engineer can, therefore, often specify the construction of those connections that are predictable and appropriate.

Connections that constrain the boundaries of a structure to a greater degree give that element increased rigidity and load-carrying capacity. For example, a beam with its ends clamped, hence restrained from rotating and displacing, will bend less under the same load than a beam free to rotate at the ends (Figures 2.1*a* and *b*). It will, therefore, take a greater load to produce a deflection on a clamped beam equal to that of the latter beam. It would seem from

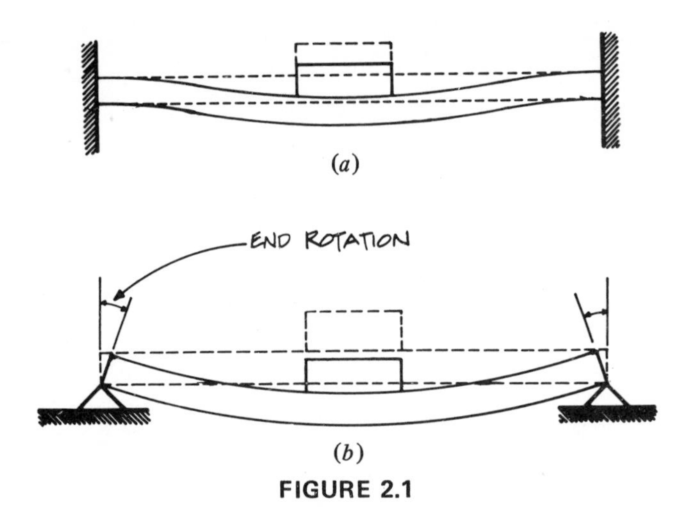

FIGURE 2.1

this example that the greatest possible degree of constraint is the most desirable for any structure; however, this is not always the case. First, stronger connections are costly as are stronger structural elements. Second, stronger connections, on occasion, cause additional stresses in the structure that are difficult to estimate. Under these circumstances, it may be less expensive to have stronger elements rather than stronger connections.

In order to understand when a greater degree of constraint produces undesirable stresses in a structure, we first observe a portal frame made by two independent posts and a lintel and then a similar portal with monolithic joints (Figures 2.2*a* and *b*). We assume that the top of the left column has moved downward for one or more reasons: (1) The left column is in the shade, the right column in the sun, so that the left column shrinks; (2) the foundation of the left column has settled; (3) the curing of the concrete in the left column has not been the same as that in the right column and has caused more shrinkage.

In any case, as the top of the left column moves downward, the posts and lintel that are less constrained do not bend, while the columns and beam in the other frame do bend (Figures 2.3*a* and *b*). Hence, the overconstrained structure is affected by the previously mentioned distortions, while the structure with fewer constraints is not.

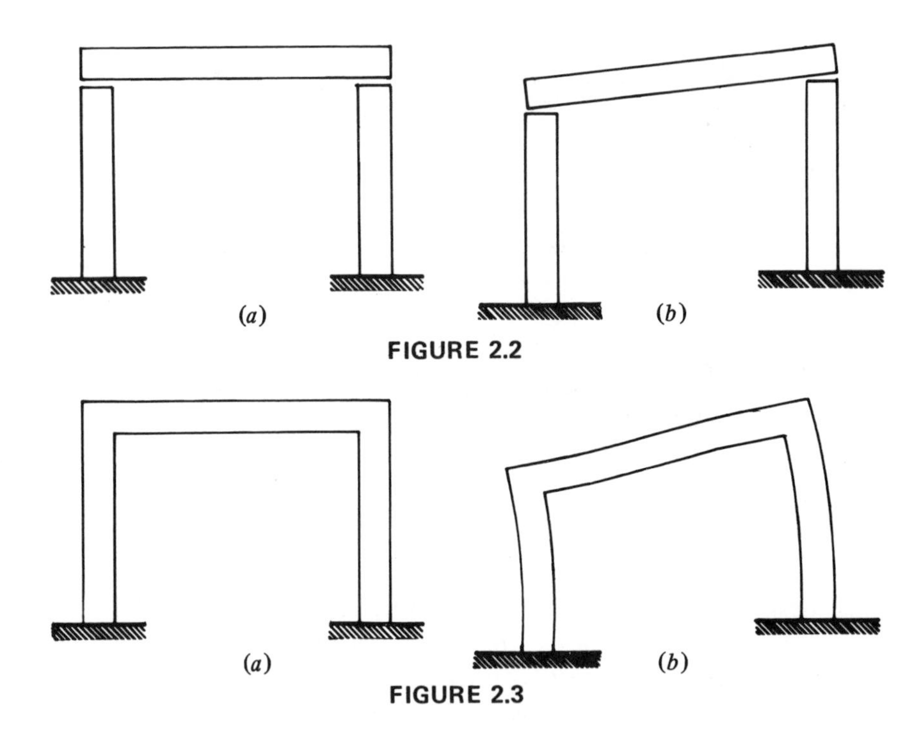

FIGURE 2.2

FIGURE 2.3

The same effect is produced by another kind of distortion called creep. Creep is a permanent contraction or extension of the structural material that occurs when compression or tension forces move the molecules of material closer or farther apart. It can be noticed, for example, in the gradual slackening of a stretched string, such as in a musical instrument.

To stress further the importance of recognizing the exact degree to which a structure is constrained, as well as the effect on structural behavior of some constraints rather than others, we consider a frame built by pin-connecting several columns and beams (Figure 2.4*a*). The frame is prone to collapse due to a lack of sufficient constraints (Figures 2.4*a* and *b*). It can be seen by inspection that placing a link in each bay of the frame, except the ground floor bay, will keep the frame from collapsing (Figures 2.5*a* and *b*). If, however, strong horizontal earthquake or hurricane loads materialize against the building, the failure of one link will create a "soft story," which will collapse. Increasing the number of links (Figure 2.6) increases the cost but

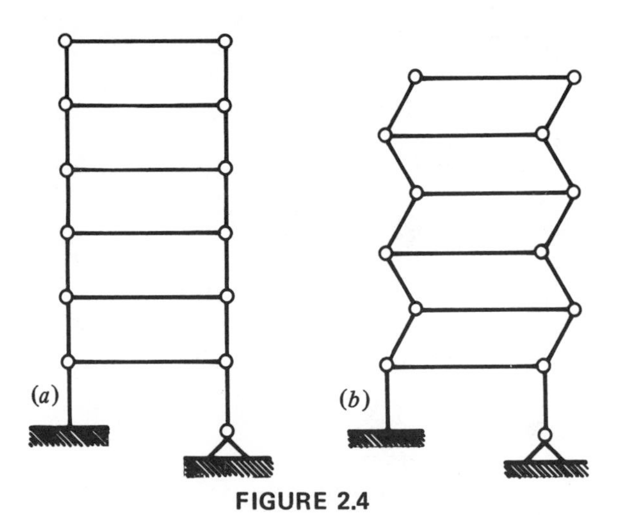

FIGURE 2.4

also the building's reserve of strength. In fact, six links of the frame in Figure 2.6 must fail before collapse occurs. In the process of overcoming the resistance of six links, a considerable part of the earthquake's energy will be spent, so that the surviving bars will not be tested by the full force of the quake.

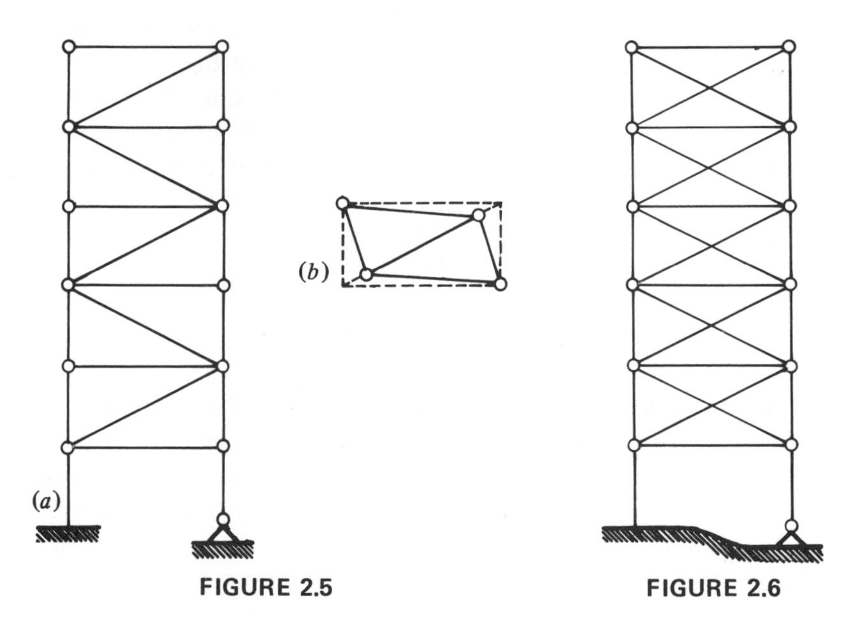

FIGURE 2.5

FIGURE 2.6

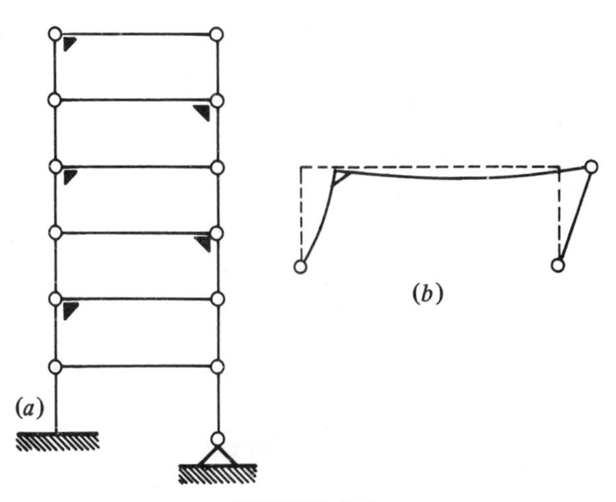

FIGURE 2.7

The same frame in Figure 2.4 can be stabilized by using at least five clamps or wedges, each of which will freeze one corner of each bay to 90° and, therefore, the other three corners of the bay as well (Figure 2.7*a*). The building in Figure 2.7*a* can be as strong as that in Figure 2.5. The latter frame, however, has greater rigidity: Its bays will deform only if their bars extend or contract, as shown in Figure 2.5*b*, while the bays in Figure 2.7*a* will deform when their bars bend (Figure 2.7*b*). It can be demonstrated on a small wooden or steel bar how much easier it is to bend a structural element than to stretch it.

Depending on design criteria, the architect-engineer will choose the appropriate frame system from the two choices considered. He/she will do so only if aware of the influence of constraints on structural behavior. For example, the rigid frame in Figure 2.5 is a better structure for the comfort of occupants and the integrity of partitions, curtain walls, and other nonstructural building components. In areas of high earthquake risk, however, the rigid frame would take the full strength of the horizontal load, while the resilient frame, like a bamboo cane in the wind, would bend rather than break.

We now turn to various constraint symbols used in structural schemes, their function, practical realization in construction, and, finally, the precise extent to which a structure is constrained by the connections shown in the structural scheme. We limit our discussion

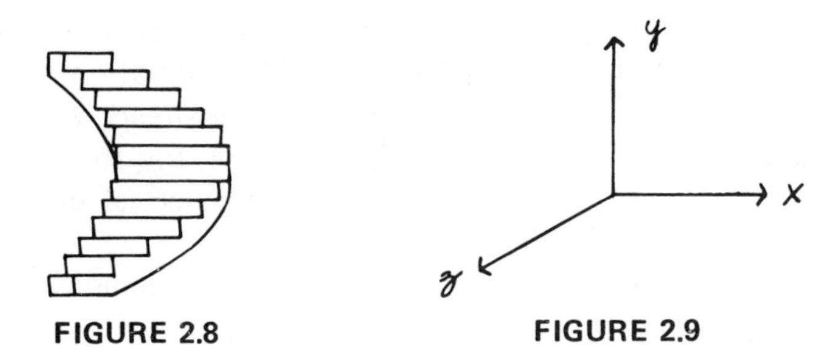

FIGURE 2.8 FIGURE 2.9

to structures and constraints in one plane, excluding those structures that have their geometries and loads projected into space (Figure 2.8).

On a Cartesian plane such as the xy-plane (Figure 2.9), a structural element has three degrees of freedom: It can have displacements parallel to the x- and y-axes and rotation around the z-axis (clockwise from y to x; counterclockwise from x to y). Plane constraints can, therefore, be classified into three groups: simple, double, and triple constraints, all depending on the number of degrees of freedom suppressed by a given constraint. Simple constraints prevent only one movement of the section of the structure to which they are applied. If the structure is rigid, the same movement is prevented throughout the entire structure.

The three equal constraints shown in Figure 2.10 are called rollers. The symbol used for a roller suggests that it allows translation on the plane of rolling and rotation around the pin at its top, while preventing any translation perpendicular to the plane of rolling. This means that the roller is able to provide an adequate reaction, a force of opposite sign but equal magnitude and direction, for any load per-

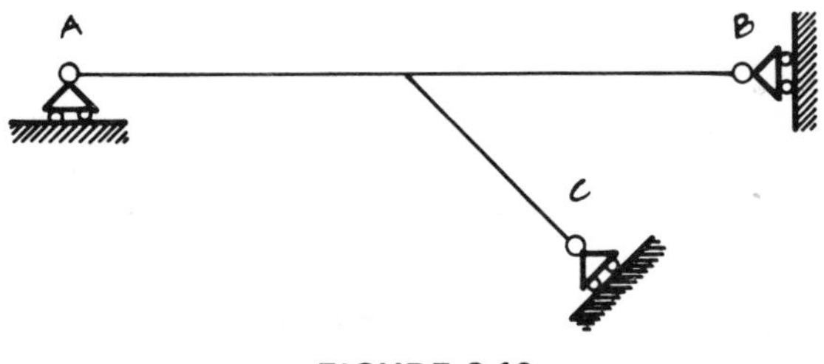

FIGURE 2.10

pendicular to the plane of rolling. The reactive force will neutralize the load, thus avoiding displacement. Since translation along the plane of rolling is permitted, the roller does not react with a force along this direction. Since rotation is also free, the roller does not react to any cause for rotation (moment) with an equal and opposite moment.

To summarize, the roller prevents translation perpendicular to the plane of rolling and reacts with a force having that direction. It allows translation parallel to the plane of rolling and does not provide a reaction in that direction. It allows rotation and does not provide a reactive moment.

In general,

a constraint reacts with forces and moments similar to the type of movements it prevents.

The structure in Figure 2.10 is constrained by three rollers: The roller at A prevents vertical translation and reacts with a vertical force; the roller at B prevents horizontal translation and reacts horizontally; the roller at C prevents any translation perpendicular to a plane inclined with the angle α and reacts with a force perpendicular to that plane.

In construction practice, a roller can be made in several ways, depending on the material used for the constraint itself and the constrained structure. For a concrete beam, for instance, roller behavior can be realized by using a neoprene pad (Figures 2.11*a* and *b*). In fact, after the initial vertical shrinkage of the pad due to the dead weight of the beam, there will be practically no additional vertical displacements. But if any cause for horizontal translation (temperature variations or breaking forces on bridge beams) tends to displace the beam end horizontally, the pad will allow such translation, due to its low transversal rigidity. If the beam tends to bend under the action of live loads, pressure exerted by the underside of the beam will not be evenly distributed on the pad; rather, it will be greater on one side and lower on the opposite side, causing the upper face of the pad to rotate with respect to the lower. This rotation will permit an equal rotation of the constrained section of the beam (Figure 2.11*b*).

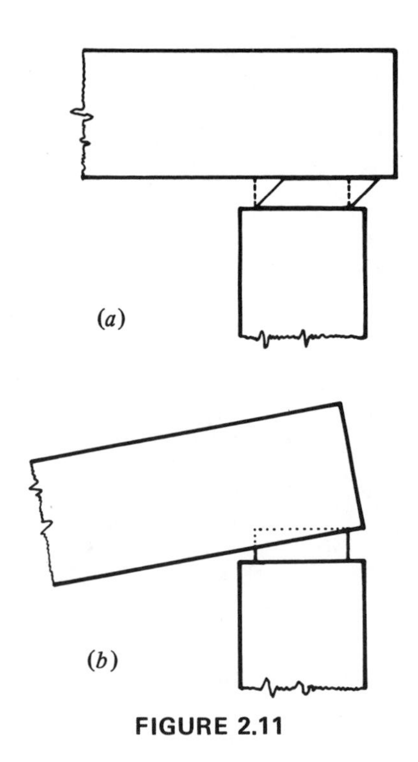

FIGURE 2.11

A steel roller may be constructed as shown in Figures 2.12*a* or *b*, which reveal the roller's behavior. Since movements of structures are usually small, the steel cylinder in Figure 2.12*a* never rotates many degrees. Hence, the unused part can be sliced in order to save material (Figure 2.12*b*).

A link or pendulum is a constraint that behaves exactly like a

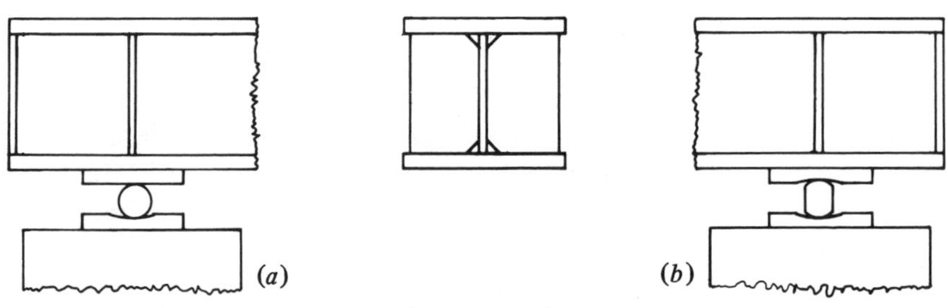

FIGURE 2.12

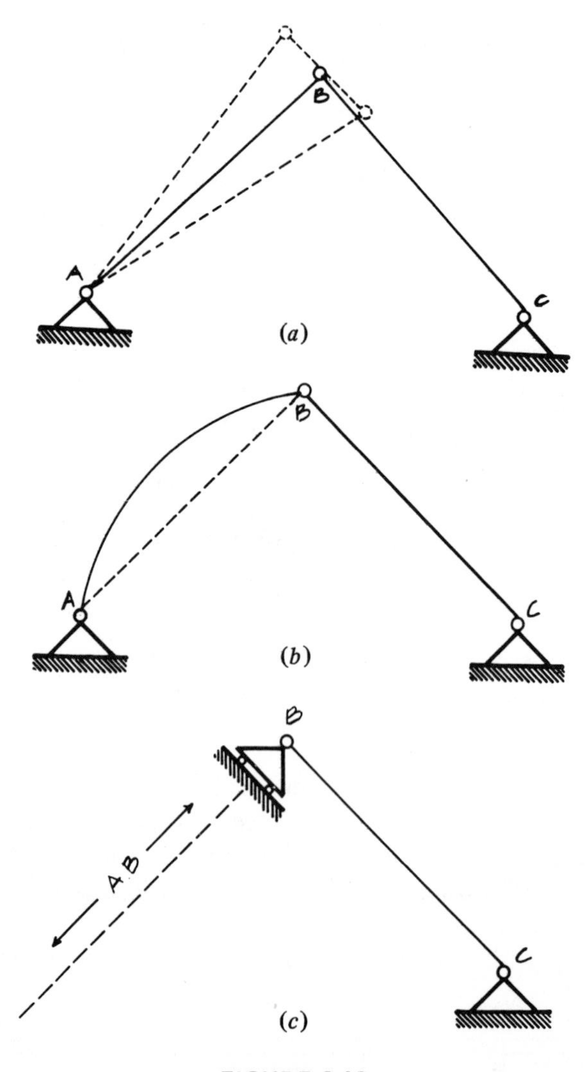

FIGURE 2.13

roller with its plane of rolling perpendicular to the direction of the link; however, it is constructed differently. A link is, in fact, a straight or curved bar having terminal hinges (definition follows), while a roller is a pointlike constraint. For example, in Figure 2.13*a*, link *AB* is equivalent to a roller that constrains motion in the direction *AB* and permits displacements perpendicular to *AB*. Similarly, in Figure 2.13*b*, the arched element *AB* behaves like a roller. Figure

2.13*c* shows the equivalent roller that prevents displacement along the dotted line *AB* in Figure 2.13*b*.

The constraints shown at points *A* and *C* in Figures 2.13*a* and *b* are called external or terminal hinges. Once again, the symbol used gives an idea of the function of this constraint. The end of a bar constrained by a hinge may rotate around the pin, but the pin is fixed to the ground, and there can be no vertical or horizontal translations of the bar end. The hinge is, therefore, a double constraint. In accordance with the general rule, a hinge reacts with two forces along the directions of displacements that it prevents. Since rotation of the constrained section is permitted, the hinge does not provide a reactive moment.

The constraint shown at point *B* in Figures 2.13*a* and *b* is called an internal hinge. It is also a double constraint, since it compels end *B* of bar *AB* to follow the same displacements along the x and y-axes as the terminal point *B* of bar *CB*. Rotation of the terminal section *B* of bar *CB* remains independent from that of section *B* of bar *AB*. For instance, bar *AB* may remain still while section *B* of bar *CB* rotates around the pin connection. Practical applications of external and internal hinges and of other constraints are shown in Figures 2.14–2.23.

FIGURE 2.14. View of one of the rollers supporting the roof of the George Washington Bridge Bus Terminal, New York City. Structures by Pier Luigi Nervi. Photograph by Miron Cohen.

FIGURE 2.15. An external hinge for a steel arch. Photograph by Yoram Finkelstein.

FIGURE 2.16. The fixed end of a steel arch. Photograph by Duc Mau Thai.

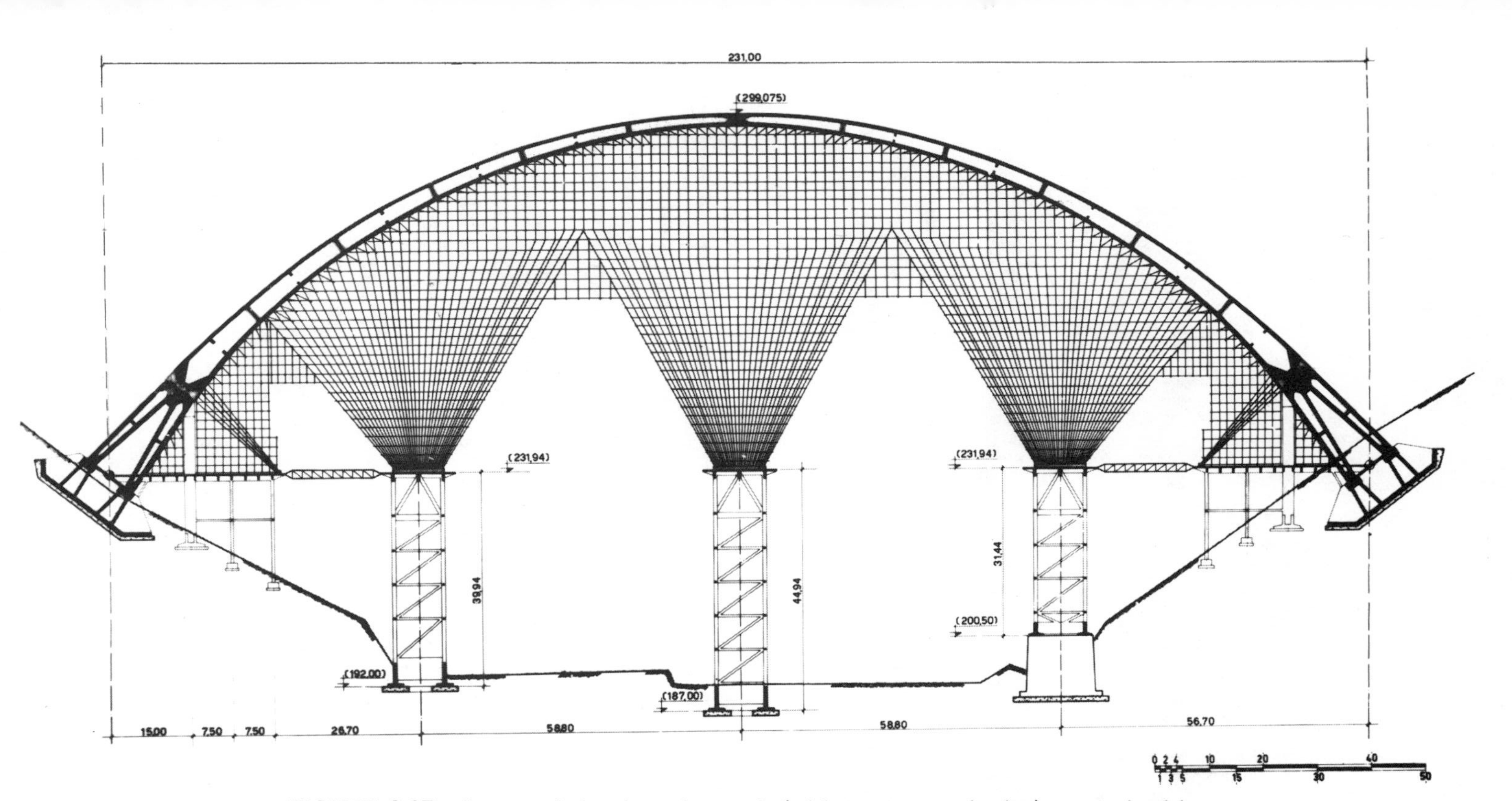

FIGURE 2.17. Structural drawing of an arch (with temporary shoring) constrained by concurrent links at its ends. Design by Riccardo Morandi for the viaduct Fiumarella in Catanzaro, Italy. Photograph courtesy of Professor Morandi.

FIGURE 2.18. The project in Figure 2.17 after completion. All the hinges of the arch were blocked after steady values were attained in the creep and shrinkage of the concrete and in the settlement of the foundation. Structure by Riccardo Morandi, Rome. Photograph courtesy of Professor Morandi.

Occasionally, two links may be used to constrain one section of a structural element. Since each link is a simple constraint, the combination results in a double constraint, which is shown in Figure 2.24*a* at points A and C. The two concurrent links constraining section A of the arch in Figure 2.24*a* are together equivalent to an external hinge placed at A'; therefore, the model structure could also be shown as in Figure 2.24*b*.

If the concurrent links connect two separate parts of the structural axis, they are equivalent to an intermediate hinge located at their point of concurrence (Figure 2.24*c*). The two parallel links constraining section C of the arch on Figures 2.24*a* and *b* prevent translation of section C in the direction of the links unless they shrink or expand, which is ruled out since *constraints are assumed to be infinitely rigid*. According to the general rule, the two parallel links exert a reaction in the direction of their axes.

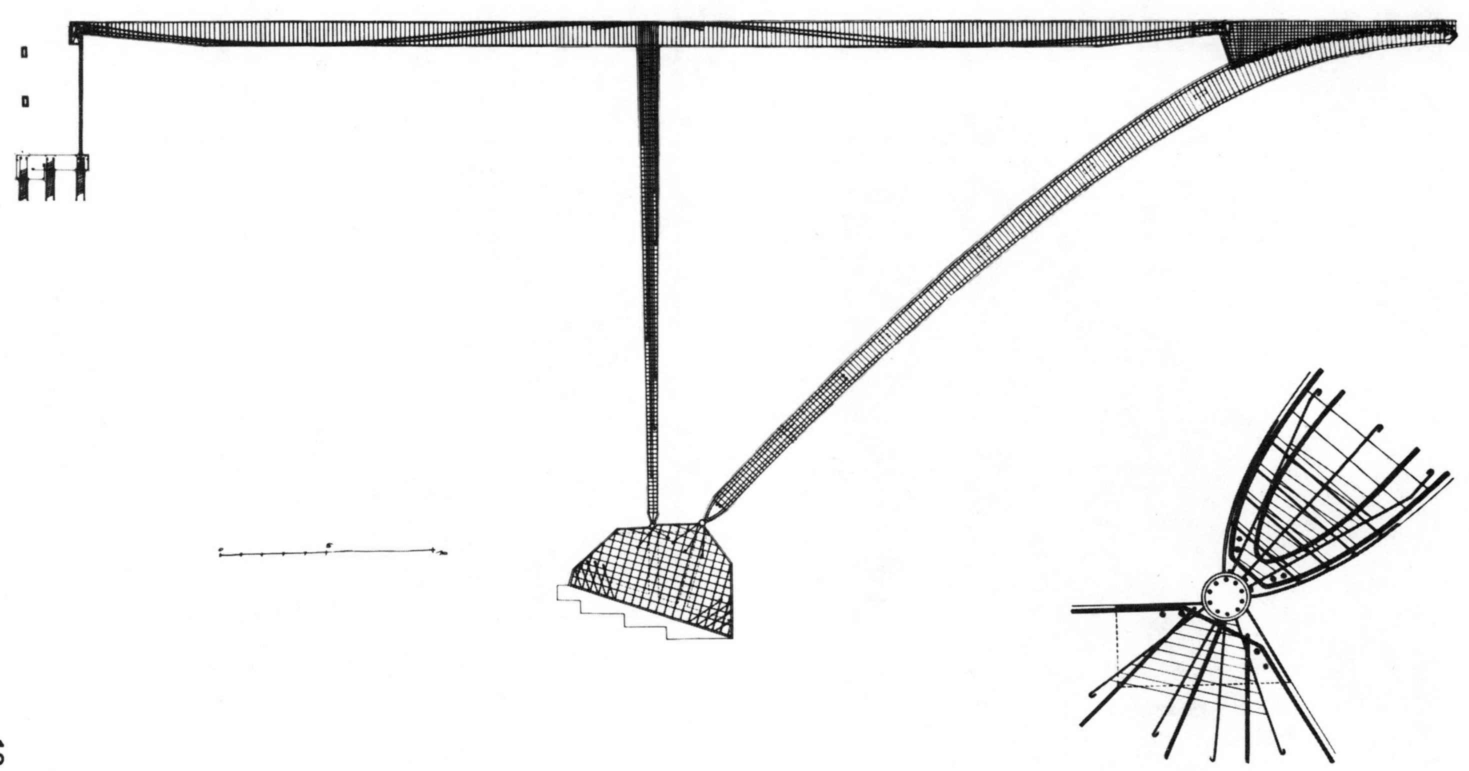

FIGURE 2.19. Structural drawing with detail of the external hinge of a concrete bridge. Structures by Riccardo Morandi, Rome. Photograph courtesy of Professor Morandi.

FIGURE 2.20. View of the external hinges of the concrete bridge in Figure 2.19. Structures by Riccardo Morandi, Rome. Photograph courtesy of Professor Morandi.

FIGURE 2.21. The concrete bridge on the river Lussia, Italy, which is constrained by the external hinges shown in detail in Figures 2.19 and 2.20. Structure by Riccardo Morandi, Rome. Photograph courtesy of Professor Morandi.

FIGURE 2.22. An internal hinge connecting concrete beams. Structures of the Magliana Bridge by Riccardo Morandi, Rome. Photograph courtesy of Professor Morandi.

Assuming small movements (so that an arc of a circle may be considered coincident with a segment of its tangent), two parallel links permit displacement perpendicular to their direction without challenging the assumption of rigidity (Figure 2.25*a*). Therefore, a reaction perpendicular to the links cannot develop, and rotation of the constrained section is prevented (Figure 2.25*b*). If, in fact, the section rotates with an angle α, then the upper link shrinks while the lower link expands. Since this has been excluded, the constraint reacts to any tendency to rotate with a moment.

FIGURE 2.23. Detail of one part of the internal hinge in Figure 2.22. Structures by Riccardo Morandi, Rome. Photograph courtesy of Professor Morandi.

This type of double constraint may be intermediate as well as terminal. It is called a slide because it permits only a sliding motion perpendicular to the links. When placed between adjacent pieces of the structure, a slide prevents the connected elements from rotating with respect to one another. One practical realization of an internal slide is shown in Figure 2.26. When the right part of the structure tends to rotate clockwise with respect to the left part, the right-end cable slackens, and the roller and the left-end cable work as compression and tension links, respectively. When the right part tends to rotate counterclockwise, the right-end cable and the roller work as tension and compression links, respectively, and the left-end cable slackens. A practical realization of an intermediate external slide is shown in this chapter's opening photograph and in Figures 2.27*a–c*.

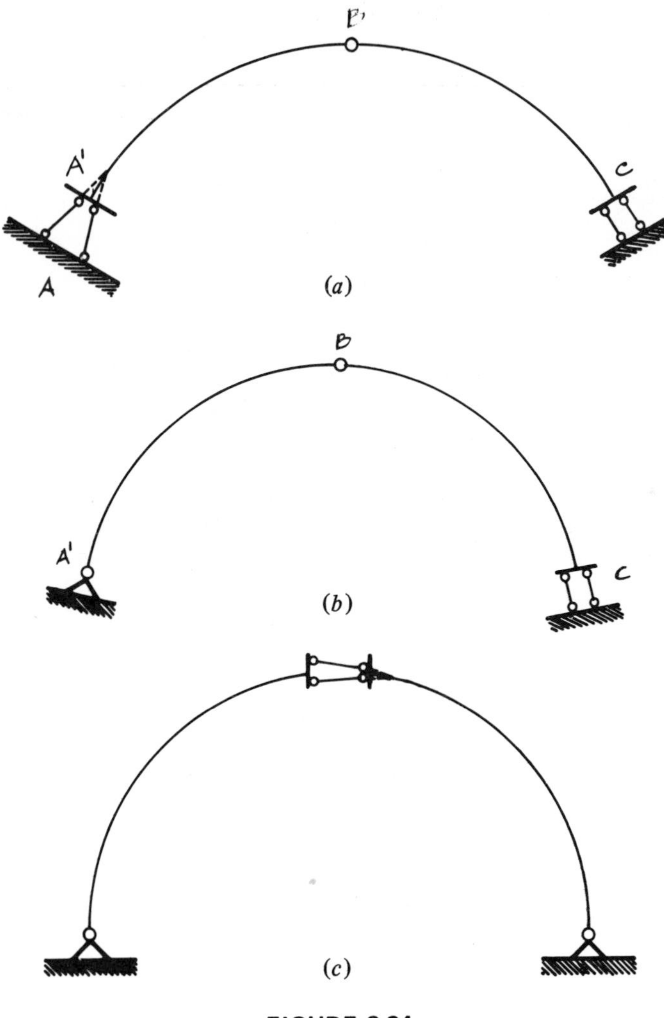

FIGURE 2.24

An intermediate (nonterminal) constraint may connect two separate bars of the same structure, in which case it is called internal, or it may connect a section of a continuous bar to the ground or to other external structures, in which case it is called external.

A very rigid beam or truss, as in Figure 2.27*a*, constrains the rotation of the columns while allowing a reduced deflection at the eleva-

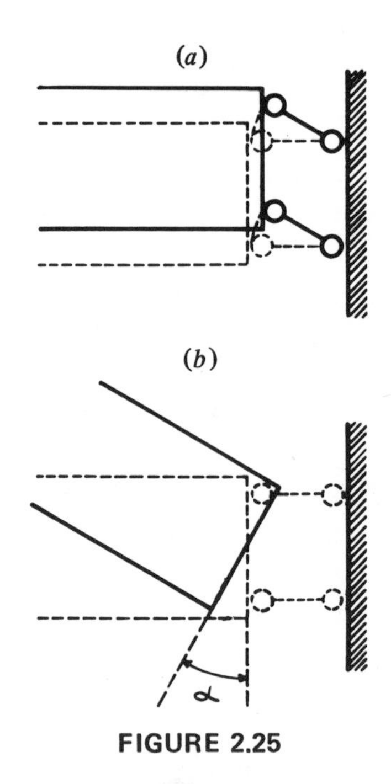

FIGURE 2.25

tion of the beam. Indeed, the rotations α of the beam ends shown in Figure 2.27*b* must occur if column sections connected to the beam are to rotate with the angle α and welding between beam and columns is not to fail.

A beam that is too rigid to deflect, as in Figure 2.27*b*, therefore prevents the rotation α for the columns as well, and column deformation (Figure 2.27*a*) appears similar to that permitted by the constraints in Figure 2.27*c*. The symbol generally used to show a triple

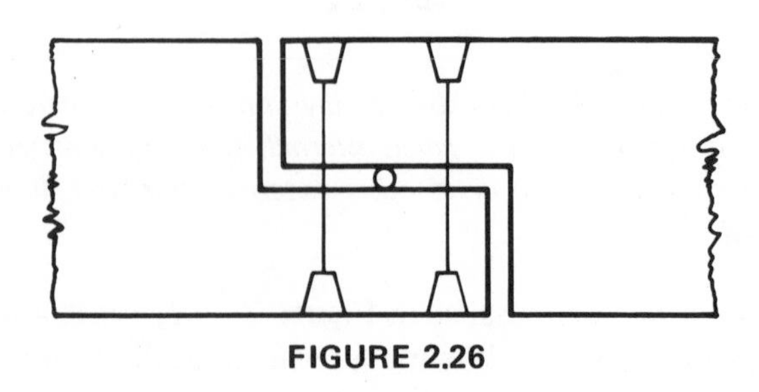

FIGURE 2.26

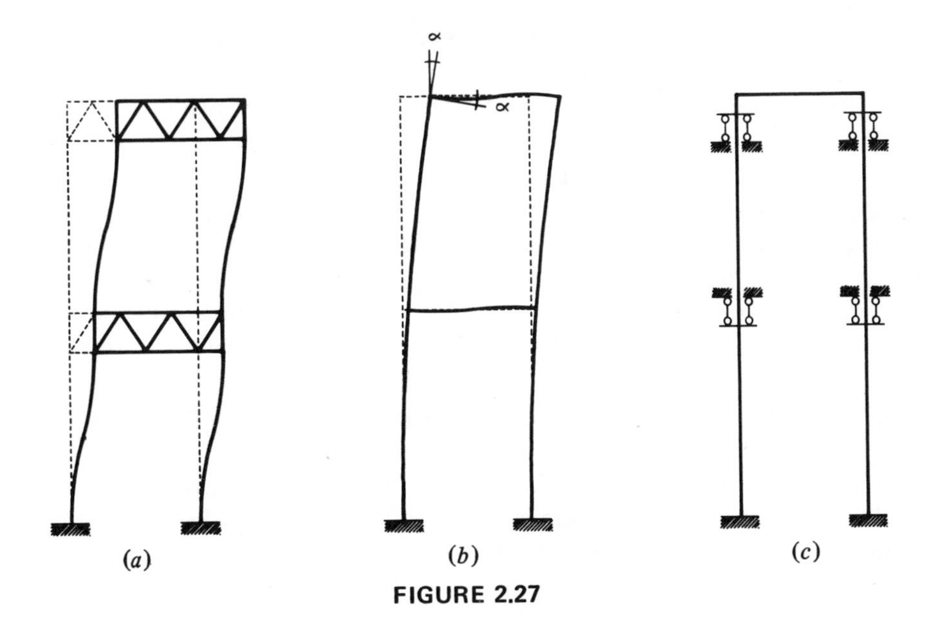

FIGURE 2.27

constraint (Figure 2.28) indicates that the constrained end is prevented from translating along either x or y and from rotating around z. Therefore, the triple constraint, also called a fixed end, reacts with forces along x and y and with a moment around z, which balances any cause of rotation. A beam built into a wall is connected to the wall by a triple constraint. In fact, it cannot translate rigidly in the horizontal or vertical direction, and it cannot swing clockwise or counterclockwise.

In reality, constraints are never perfectly rigid but are more or less deformable, plastically or elastically. A clay foundation on which a

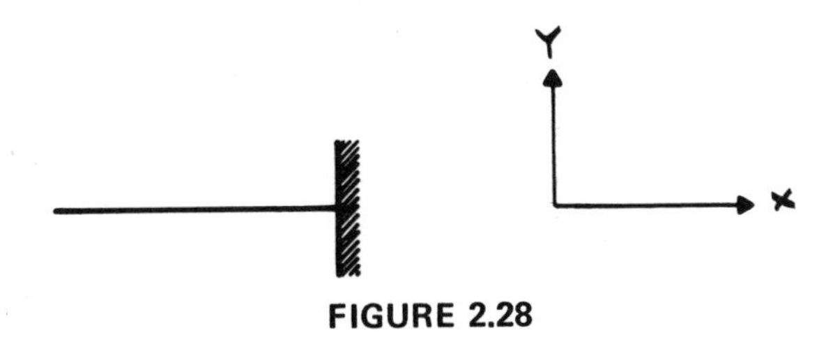

FIGURE 2.28

building is erected will slump, hopefully within acceptable limits, and the deformation will be permanent or plastic. It will not disappear if loads in the building are reduced. A well-known case is the leaning tower of Pisa in Italy.

The constrained section of a balcony slab is not, in reality, perfectly fixed, but, instead, it rotates with an angle equal to the twist of the supporting beam. If the loads are removed, deformation of the constraint (twist of the beam) disappears (Figure 2.29). This is a case of elastic constraint deformation. The following table summarizes the various types of constraints that we have been discussing. The key is a Cartesian, horizontal-vertical (x, y) frame of reference. The preceding photographs of link and hinge connections show to what extent awareness of constraint behavior can be used by a designer to improve the efficiency and elegance of structures. Mies van der Rohe's expression, "God is in the details," applies to connections as well.

Symbol						
Type	Simple	Simple	Double	Double	Double	Triple
Prevented Movements	δ_y	δ_y	δ_x, δ_y	δ_x, δ_y	δ_x, ϕ	δ_x, δ_y, ϕ
Reactions	R_y	R_y	R_x, R_y	R_x, R_y	R_x, M	R_x, R_y, M
Allowed Movements	δ_x, ϕ	δ_x, ϕ	ϕ	ϕ	δ_y	none

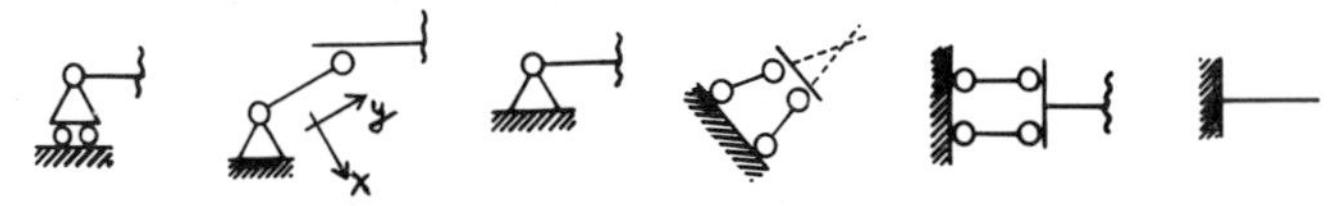

With an understanding of the behavior of the various constraints, it is now possible to *evaluate the degree of stability of plane structures*. Considering initially only external constraints, we will discuss the following points: (1) Each fixed end eliminates three degrees of freedom (x and y translations and rotation); therefore, if a structure has f fixed ends, $3f$ degrees of freedom are suppressed. (2) Each external hinge and each external slide eliminates two degrees of freedom; therefore, if a structure has h external hinges and slides, $2h$ degrees of freedom are suppressed. (3) Each roller and each link eliminates one degree of freedom (translation perpendicular to plane of rolling

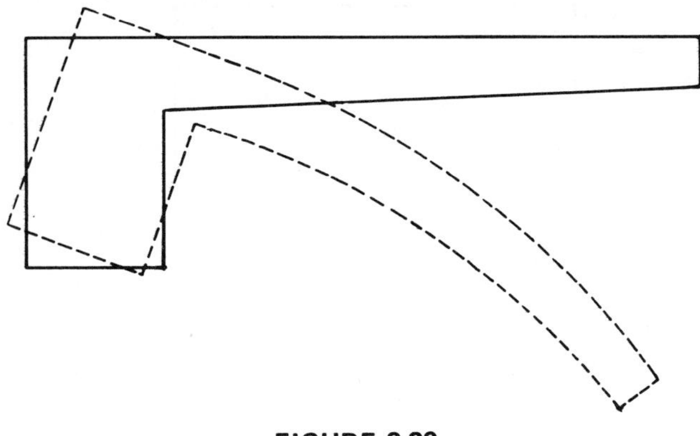

FIGURE 2.29

or translation parallel to the link); therefore, if a structure has s simple constraints, s degrees of freedom are suppressed. Hence, the total number of degrees of freedom suppressed by external constraints can be expressed as:

$$3f + 2h + s.$$

If the structure has n monolithic parts, each of which has three degrees of freedom (x and y translation plus rotation), then $3n$ degrees of freedom must be suppressed to prevent rigid motion of the structure. The necessary condition for the stability of a structure is, therefore,

$$3f + 2h + s = 3n.$$

If $3f + 2h + s < 3n$, then the structure is unstable: It can move, or part of it can move ridigly. If $3f + 2f + s > 3n$, then the structure is overconstrained. It is important to recognize that these conditions are not sufficient to define completely the degree of stability of a structure.

The portal in Figure 2.30, for instance, has three constraints and three degrees of freedom. In fact, $h = 1, s = 1$, and $n = 1$; hence,

$$3f + 2h + 1s = 3n.$$

The portal, however, is unstable under the action of horizontal wind load.

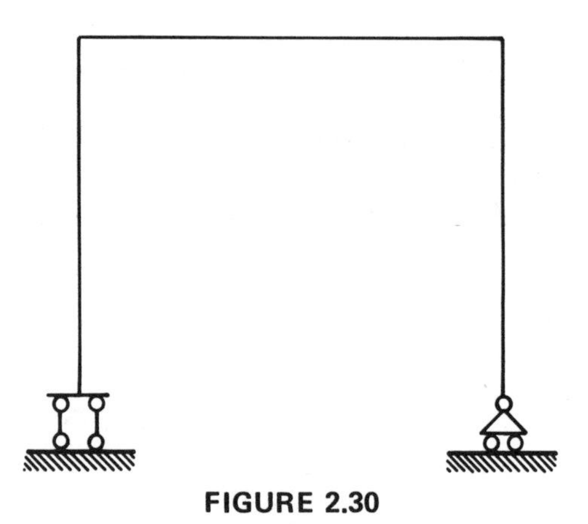

FIGURE 2.30

Example: The structure in Figure 2.31 has two monolithic parts ($n = 2$); two simple constraints, the roller and the link ($s = 2$); two double constraints, the hinge and the slide ($h = 2$); and one fixed end ($f = 1$). Then,

$$3f + 2h + s = 3(1) + 2(2) + 2 = 9$$

and

$$3(n) = 3(2) = 6.$$

With nine constraints and six degrees of freedom, the structure is threefold overconstrained. We now extend the same reasoning to internal constraints.

Each internal hinge connecting b number of bar ends compels all bar ends to have equal x and y displacements while leaving their rotations independent. It could be said that one bar end taken as the leader is free to move at will, while the remaining $(b - 1)$ bar ends must follow the x and y displacements of the leader. Since $(b - 1)$ bar ends have each lost two degrees of freedom, the internal hinge connection provides $2(b - 1)$ constraints.

Example: Ends 2 and 3 of the structure in Figure 2.32*a* can deflect independently from each other and the column top. After we connect the beams and column with a hinge, beam ends 2 and 3 will deflect only as much as the column shrinks. The structure in Figure 2.32*a* has $3n = 9$ degrees of freedom and $3f = 9$ constraints;

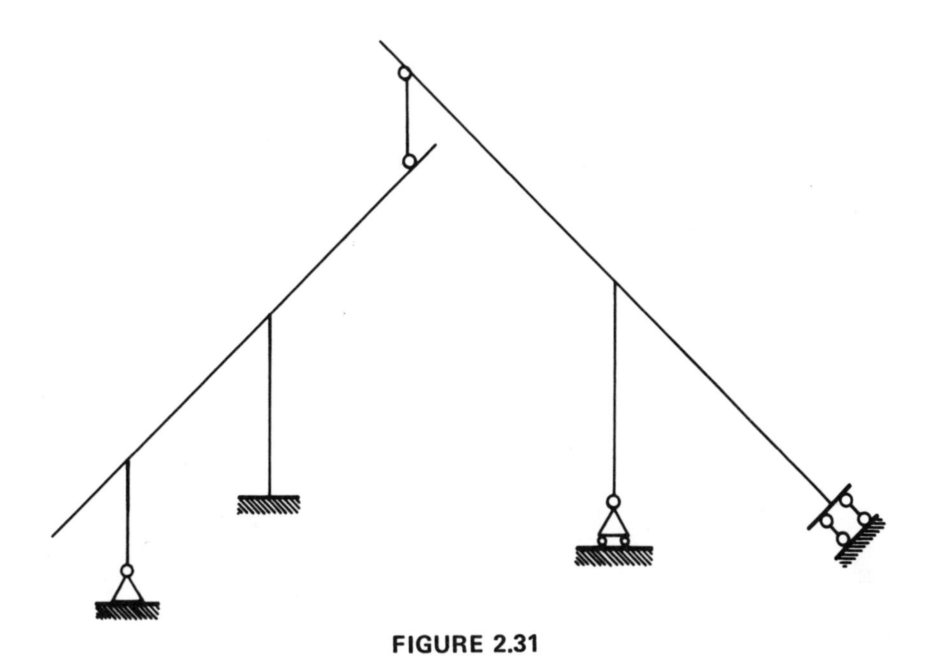

FIGURE 2.31

therefore, it is stable. The structure in Figure 2.32*b* has $2(3 - 1) = 4$ additional constraints, and it is fourfold overconstrained.

An internal hinge replacing a welded connection releases $(b - 1)$ degrees of freedom, since it returns rotational independence from the leader to $(b - 1)$ bars. An internal hinge can, therefore, be regarded as a connection or a release, depending on the vantage point. Everything that has been said about the internal hinge is valid for the internal slide, which is a double constraint as well. Of course, an internal slide compels $(b - 1)$ bar ends to follow the rotation and one displacement of the leading bar end.

Each internal welded connection joining b number of bars of a structure suppresses $3(b - 1)$ degrees of freedom in the structure or provides $3(b - 1)$ constraints. In fact, when the end of the leading bar moves to the new plane coordinates x and y and rotates with angle ϕ, the ends of the $(b - 1)$ following bars are constrained to move to the same x and y positions and rotate with the angle ϕ, each thereby losing three degrees of freedom. In total, the structure loses $3(b - 1)$ degrees of freedom after the bars are welded together.

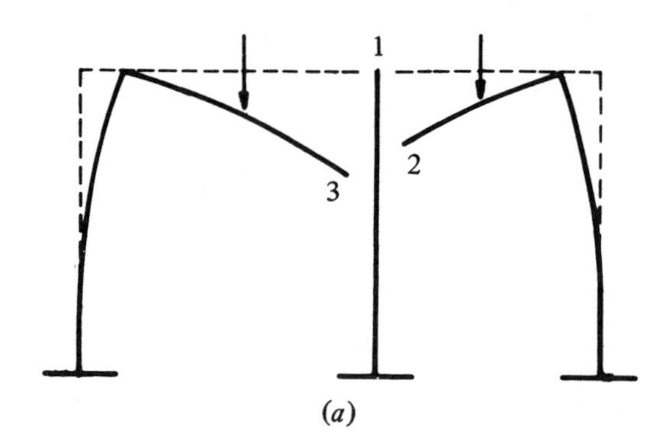

(a)

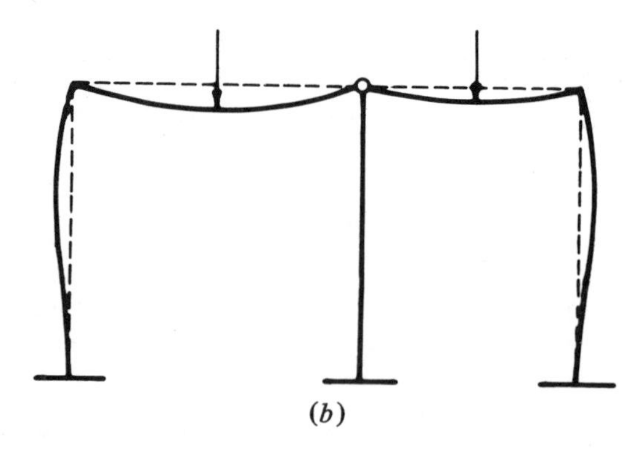

(b)

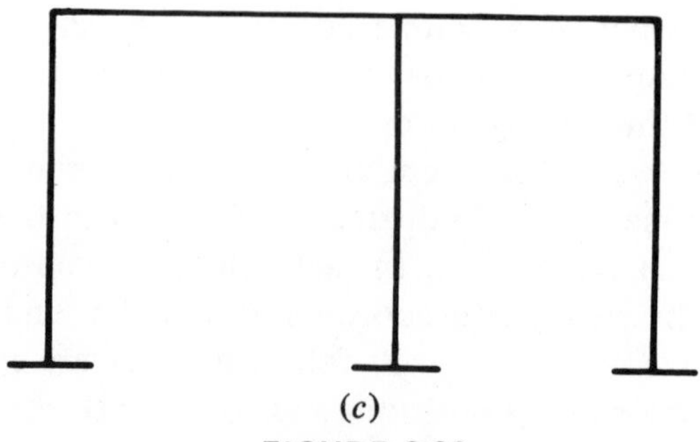

(c)

FIGURE 2.32

Example: If the beams of the structure in Fig. 2.32a are welded to one another and to the column, then $3(3 - 1) = 6$ additional constraints are introduced, and the structure becomes sixfold overconstrained.

In evaluating the degree of stability of a structure, monolithic or welded joints can be disregarded and all monolithically connected bars in the structure considered as one element having three degrees of freedom. It is essential, however, that the bars of the structure do not form closed loops. Otherwise, welded connections must be taken into account to obtain the true degree of stability of the structure.

Example: The monolithic frame in Figure 2.32c can be considered as a single element ($n = 1$) having $3n = 3$ degrees of freedom and $3f = 9$ constraints; hence, sixfold overconstrained.

Taking into account the connection between beams and column does not change the result. The connection, in fact, joins the 3 bars in Figure 2.32a, which have $3n = 9$ degrees of freedom, and it adds $3(3 - 1) = 6$ internal constraints to the 9 external constraints. In total, there are $9 + 6 = 15$ constraints and 9 degrees of freedom; the difference is, again, 6.

The frame in Figure 2.33a has a closed loop, the second floor bay. If we consider it as a single element with three degrees of freedom, we come to the conclusion that the frame is sixfold overconstrained. If we consider that at point P a welded connection joins the four bar ends in Figure 2.33b, we say that $3(4 - 1) = 9$ internal constraints are added to the nine external ones, for a total of 18. Since the three monolithic elements have $3n = 9$ degrees of freedom, we reach a different conclusion: The frame is ninefold overconstrained.

The first conclusion is false because it results from considering a monolithic element that includes a closed bay. The second conclusion is correct; to prove it, we consider an additional welded joint at point Q. We therefore add the $3(2 - 1) = 3$ internal constraints of joint Q to the nine internal constraints of joint P and the nine external constraints for a total of 21. The four monolithic elements

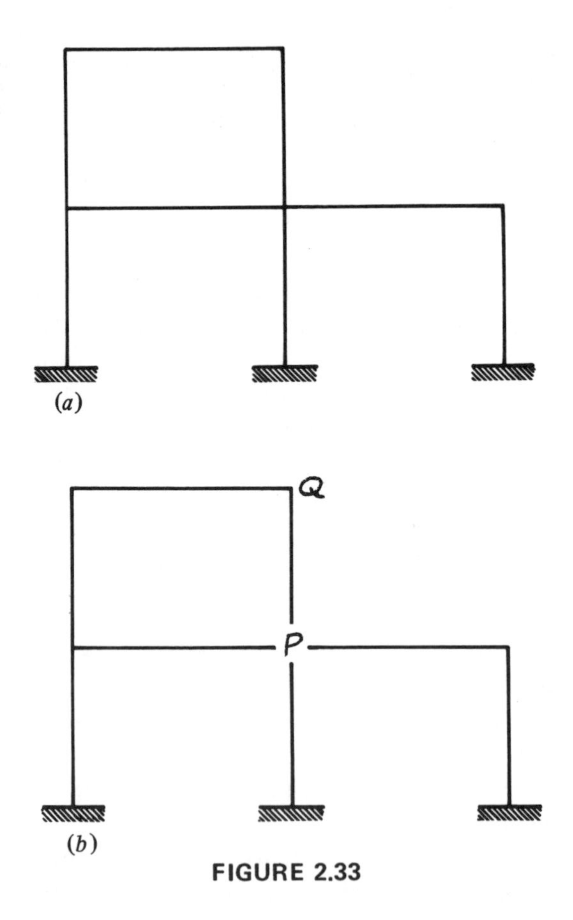

FIGURE 2.33

have $3n = 12$ degrees of freedom, and the conclusion is, again, that the frame is ninefold overconstrained.

2.3. EXTERNAL FORCES: LOADS OF THE MODEL STRUCTURE

The schematization of the real load distribution on the structure concludes the description of the model structure. Real loads can be represented symbolically by concentrated forces or various distribution diagrams.

Examples: A beam load of constant intensity g (gravity) is represented by a rectangle (Figure 2.34*a*) having its base along the beam axis and its height labeled g to indicate the intensity of the load. The total load, resultant of the load distribution, equals the product of the intensity g times the length l of the beam axis. The total load thus equals the area of the rectangle, and it is ideally placed at the intersection of the diagonals, which is the center of the rectangle.

A load of books filling only half of a set of shelves on one side of a diagonal is represented by a triangle in Figure 2.34*b*. Base l of the triangle represents the extension of the shelves on the floor, and the height g_{max} of the triangle represents the weight of the books stacked at the full end of the shelves. The book load decreases linearly from g_{max} to zero at the empty end of the shelves. The total load equals the product $(l/2)\, g_{max}$ or the average load $g_{max}/2$ times the extension of the load l, and it is ideally applied at the center of

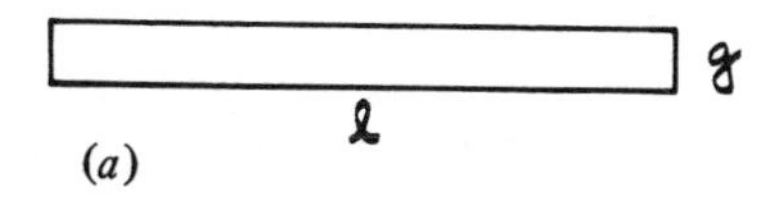

(*a*)

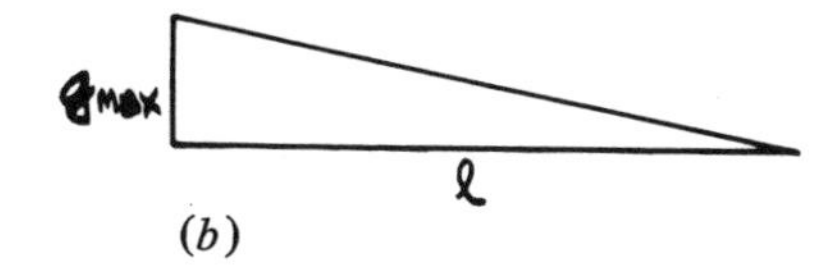

(*b*)

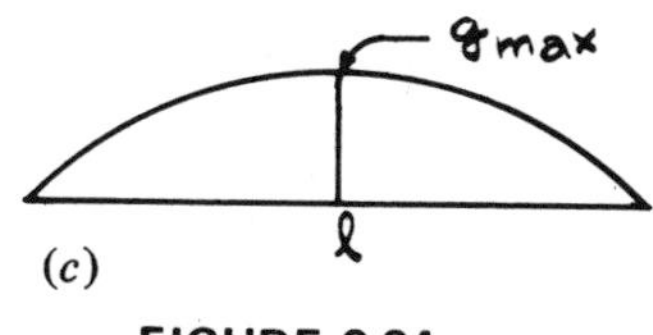

(*c*)

FIGURE 2.34

the triangle, which is at a distance $l/3$ from the left end of the triangle; $(l/2)\,g_{max}$ is also the area of the triangle.

A different arrangement of books on the shelves would be represented by a parabolic load diagram of intensity g_{max} at the center (Figure 2.34*c*). In that case, the total load would be $(2/3)l\,g_{max}$, which is the area of the parabola of rise g_{max} and base l, and it is ideally placed at the center of the shelves. The resultant load can, on occasion, be conveniently used instead of the diagrammatic distribution.

Real loads on a structure are usually divided into three kinds: dead loads, for which the abbreviation DL is used; superposed dead loads, with the abbreviation SDL; live loads, abbreviated as LL.

Dead loads are the gravity loads of structural components. They can, therefore, be very accurately estimated in intensity and distribution, providing that sizes and materials of structural members have been specified.

Superposed dead loads are also permanent, but they are derived from the weight of such nonstructural components as flooring, ceilings, partitions, lighting fixtures, heating and cooling ductwork, and piping. These loads can also be closely estimated, but not so precisely as the dead loads.

Live loads are transient and include the weight of furnishings, people, snow, wind, and other forces. Any list of live loads would include at least wind pressure on the windward side of buildings and the depression on the leeward side; the uplift produced by wind on roofs; the pressure of water or earth against retaining or basement walls; buoyancy forces on submerged building foundations; vertical and horizontal inertia forces due to sudden motions of the foundation during earthquakes. Live loads are, therefore, not so easy to estimate as dead and superposed dead loads. If, however, records of wind or earthquake patterns and intensity are available, statistical evaluations of these loads are possible. To eliminate individual guesswork and subjective interpretations of data, live loads are specified by the American National Standard Institute in *Building Code Requirements for Minimum Design Loads in Buildings and Other Structures, A 58.1-1972.*

The effects of live loads on structures are usually evaluated sepa-

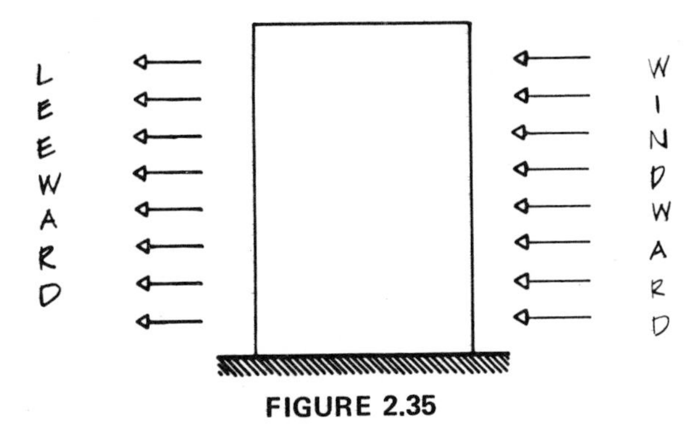

FIGURE 2.35

rately from those of dead loads. Live loads may, indeed, increase or decrease stresses and deformations produced by dead loads on some structural components. For example, uplifting wind forces neutralize part of the weight of the roof.

Indeed a theorem in fluid dynamics due to Bernoulli states that the energy of a current is a constant given by the sum of static and kinetic energy

$$\frac{p_1}{\gamma} + \frac{V_1^2}{2g} = \frac{p_2}{\gamma} + \frac{V_2^2}{2g},$$

where

p = static pressure of the fluid,
γ = density of the fluid,
V = velocity of the fluid,
g = acceleration due to gravity (approximately 32.2 ft/sec^2).

A wind current sweeping over a roof at considerable velocity V_2 has an even greater kinetic energy $V_2^2/2g$. The air below the roof has zero velocity V_1 and, therefore, great static pressure p_1, so that the lack of kinetic energy is balanced by considerable static energy. The pressure differential $p_1 - p_2$ tends to lift up the roof.

Pressure and depression produced by wind on the opposite sides of a building (Figure 2.35) decrease stresses on the building's foundation on the windward side but increase them on the leeward side. Since

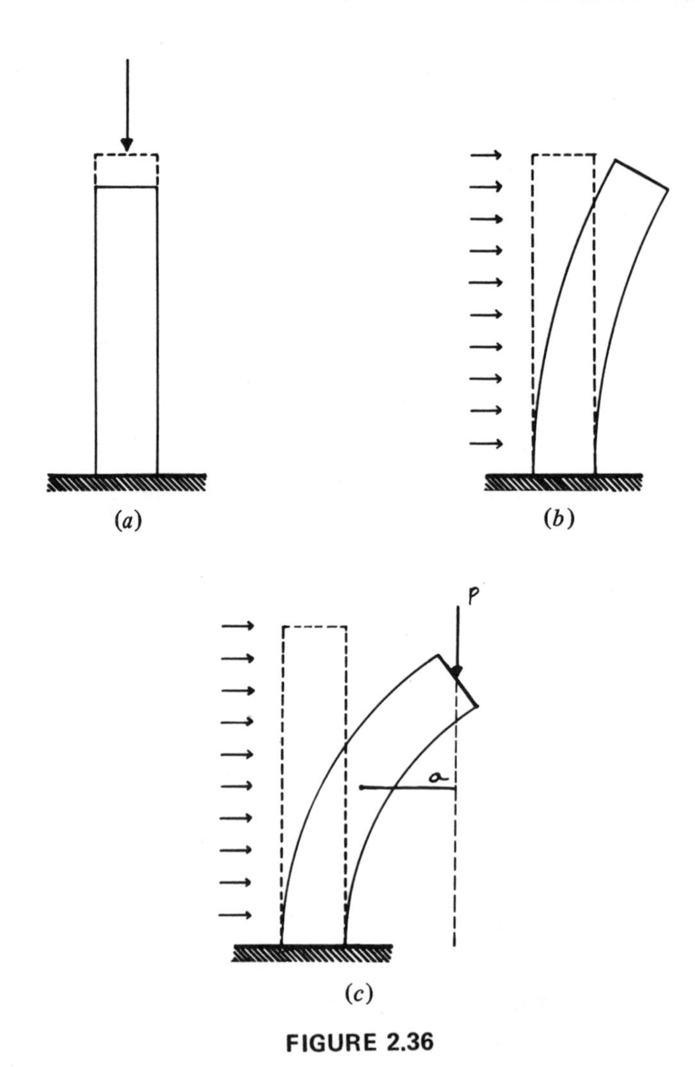

FIGURE 2.36

live loads may or may not materialize, the most dangerous conditions must be anticipated in structural design. If dead loads alone produce critical conditions for a given structural component, then the effect of live loads will be excluded in that component's design. Otherwise, the effects of dead and live loads are combined in accordance with the principle of superposition, which can be stated as follows:

The effect of a combination of loads is the sum total of the effects due to each individual load.

This principle can be applied in the design of most architectural structures; it is not, however, always valid.

Example: A column is subject to a vertical force P at its top delivered by the roof it supports. The column is also subject to a wind load against its side (Figures 2.36*a* and *b*). The right-side fibers of the column undergo a shrinkage due to the gravity force and another different shrinkage due to wind load. When wind and gravity loads are applied simultaneously, the right-side fibers undergo a total shrinkage that is the sum of the separate contractions, as long as the wind bends the column only slightly. But if the wind bends the column noticeably (Figure 2.36*c*), the lever arms of P around the various points of the column axis, such as a, are not negligible: The moments aP also produce shrinkage of the right-side fibers of the column. As a result, the contraction of the right-side fiber, due to the simultaneous application of gravity and wind loads is not the sum of the contractions due individually to P and the wind: It is, instead, a larger sum, which includes the contraction due to such moments as aP.

Example: We consider a steel bar of length L and cross-sectional area A that is subject to a traction force P_1 at one time and to a traction force P_2 at another time. The steel in the bar is characterized by a relation between the stresses P/A and the strains $\Delta L/L$ shown by the curve in Figure 2.37*a*. Under the stress P_1/A, the bar strain is $(\Delta L)_1/L$ (Figure 2.37*a*). Under the stress P_2/A, the bar strain is $(\Delta L)_2/L$.

If traction forces P_1 and P_2 are applied simultaneously to the steel bar, the stress is $(P_1 + P_2)/A$, but the strain is larger than $[(\Delta L)_1/L] + [(\Delta L)_2/L]$ (Figure 2.37*a*). If, however, total stress $(P_1 + P_2)/A$ does not exceed the limit where the stress-strain curve starts to have a variable slope, the sum of the separate strains $(\Delta L)_1/L$ and $(\Delta L)_2/L$ coincides with the combined strain $(\Delta L)_{1,2}/L$ (Figure 2.37*b*). Hence, the principle of superposition is valid in the latter case, but not in the former.

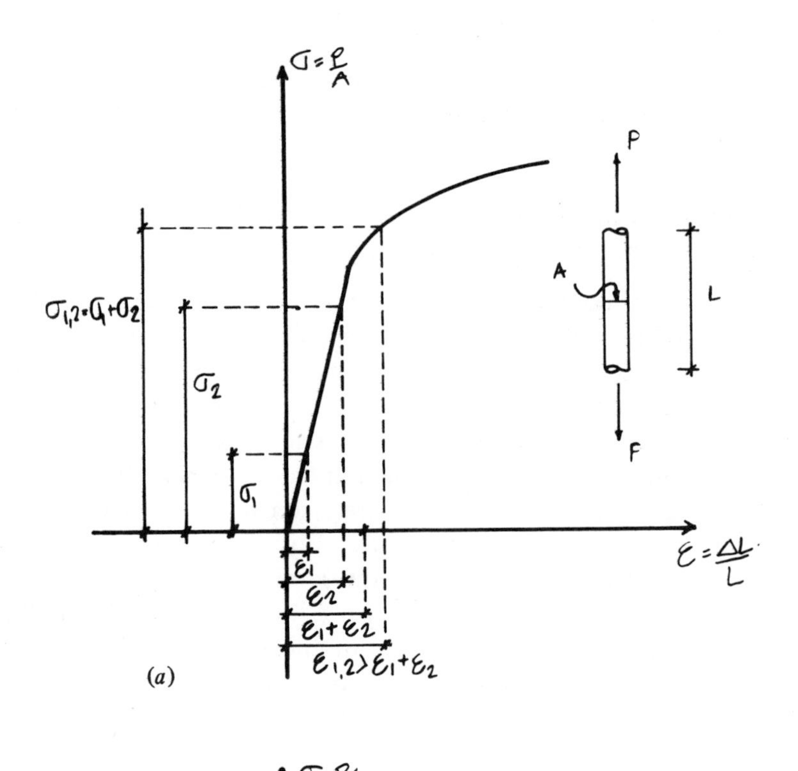

(a)

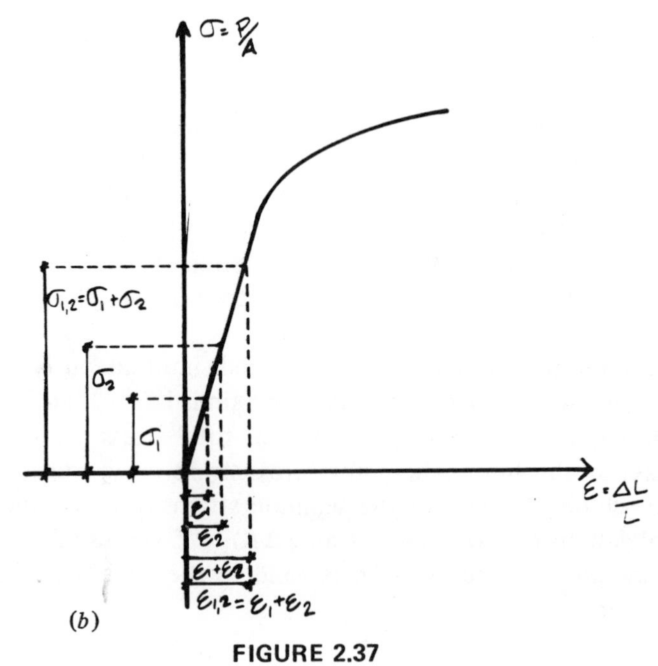

(b)

FIGURE 2.37

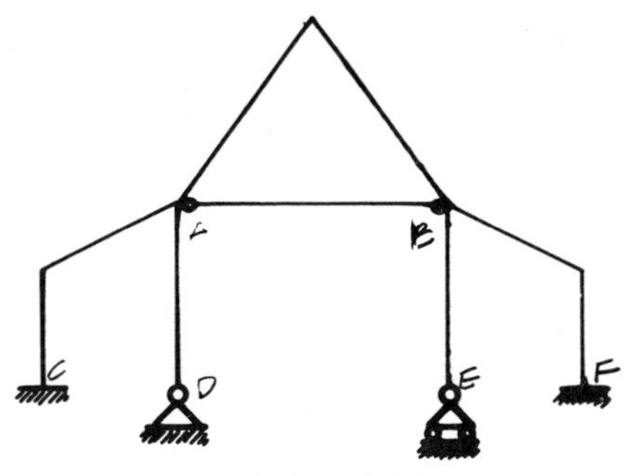

FIGURE 2.38

PROBLEMS

The following problems with solutions help clarify the method for calculating the degree of stability of structures. Problems without solutions provide practice in the method.

Evaluate the degree of stability in the following structures.

2.1. Assume AB to be a constraining link and the concrete frame to be cast monolithically (Figure 2.38).

Solution.

$$s = 2 \text{ (link } AB\text{, roller } E\text{)},$$

$$h = 1 \text{ (hinge } D\text{)},$$

$$f = 2 \text{ (fixed ends } C \text{ and } F\text{)},$$

$$3f + 2h + s = 3(2) + 2(1) + 2 = 6 + 2 + 2 = 10 \text{ (constraints)},$$

$$n = 1 \text{ (one solid structure)},$$

$$3n = 3 \text{ (degrees of freedom)},$$

$$10 > 3 \text{ (redundant structure)},$$

$$10 - 3 = 7 \text{ (redundant constraints)}.$$

2.2. Assume AB, BC, and CD to be three constraining links, and the rest of the frame to be solid (Figure 2.39). Solve.

2.3. Assume the arched frame to be solid (Figure 2.40). Solve.

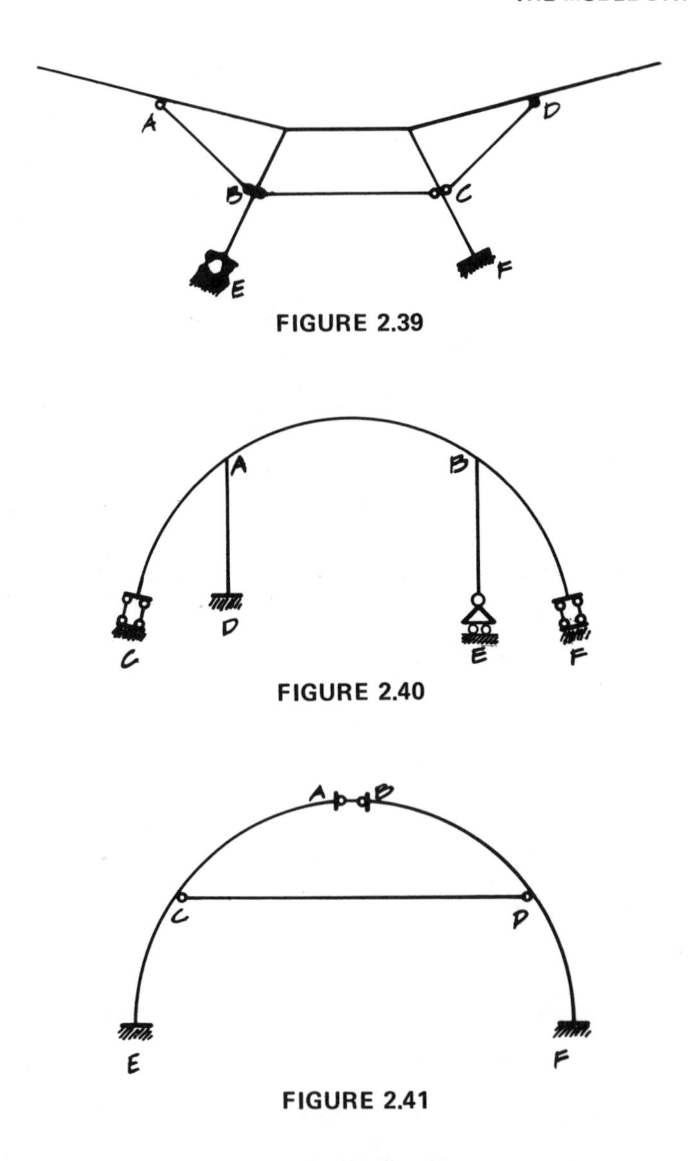

FIGURE 2.39

FIGURE 2.40

FIGURE 2.41

2.4. Assume AB and CD to be constraining links (Figure 2.41). Solve.

2.5. In Problem 2.1, do not assume AB to be a link but rather a bar hinged to the solid frame at A and B (Figure 2.38).

Solution.

$$s = 1 \text{ (roller } E\text{)},$$

$$h = 1 \text{ (hinge } D\text{)},$$

$$f = 2 \text{ (ends } C \text{ and } F\text{)},$$

$$3f + 2h + s = 3(2) + 2(1) + 1 = 9 \text{ (external constraints)}.$$

Hinge A connects bar AB and the solid frame. Hence, $2(b_A - 1) = 2(2 - 1) = 2$.

Hinge B connects bar AB and the solid frame. Hence, $2(b_B - 1) = 2(2 - 1) = 2$.

Subtotal of internal constraints: $2 + 2 = 4$.

Total of external and internal constraints: $9 + 4 = 13$.

$$n = 2 \text{ (bar } AB \text{ and solid frame)},$$

$$3n = 3(2) = 6 \text{ (degrees of freedom)},$$

$$3f + 2h + s + 2(b_A - 1) + 2(b_B - 1) = 13 > 6 \text{ (redundant structure)},$$

$$13 - 6 = 7 \text{ (redundant constraints; same as in Problem 2.1)}.$$

2.6. In Problem 2.3, assume joints A and B to be welded connections (Figure 2.40).

Solution 1.

$$s = 1 \text{ (roller } E\text{)},$$

$$h = 2 \text{ (slides } C \text{ and } F\text{)},$$

$$f = 1 \text{ (fixed end } D\text{)},$$

$$3f + 2h + s = 3 + 4 + 1 = 8 \text{ (subtotal of external constraints)}.$$

Weld A joins bars AC, AD, and AB. Hence, $3(b_A - 1) = 3(3 - 1) = 6$.

Weld B joins bars BE, BF, and BA. Hence, $3(b_B - 1) = 6$.

Subtotal of internal constraints: $6 + 6 = 12$.

Total of external and internal constraints: $8 + 12 = 20$.

$$n = 5 \text{ (bars } AC, AD, AB, BE, \text{ and } BF\text{)},$$

$$3n = 15 \text{ (degrees of freedom)},$$

$$20 > 15 \text{ (redundant structure)},$$

$$20 - 15 = 5 \text{ (redundant constraints)}.$$

Solution 2.

$$3f + 2h + s = 8 \text{ (subtotal of external constraints)}.$$

Weld A joins bar AD to bar CF. Hence, $3(b_A - 1) = 3(2 - 1) = 3$.
Weld B joins bar BE to bar CF. Hence, $3(b_B - 1) = 3(2 - 1) = 3$.
Subtotal of internal constraints: $3 + 3 = 6$.
Total of internal and external constraints: $8 + 6 = 14$.

$$n = 3 \text{ (bars } AD, BE, \text{ and } CF),$$

$$3n = 9 \text{ (degrees of freedom)},$$

$$14 > 9 \text{ (redundant structure)},$$

$$14 - 9 = 5 \text{ (redundant constraints as in Solution 1)}.$$

2.7. In Problem 2.4, do not consider CD to be a link but rather a bar hinged at C and D (Figure 2.41); solve.

2.8. The frame includes a closed loop. At least one internal weld around the loop must be taken into account, after which it is not necessary to consider additional welds. In general, only welds that form closed loops must be counted among internal constraints (Figure 2.42*a*).

Solution 1 (Figure 2.42*b*).

$$s = 1 \text{ (roller } C),$$

$$h = 1 \text{ (hinge } B),$$

$$f = 2 \text{ (ends } A \text{ and } D),$$

$$3f + 2h + s = 6 + 2 + 1 = 9 \text{ (subtotal of external constraints)}.$$

Weld E forms a closed loop by joining bars EF and EG. Hence, $3(b_E - 1) = 3(2 - 1) = 3$. After taking into account and discarding constraints A, B, C, D, and E, the frame does not have closed loops and is still made of one piece free in space. Total of internal and external constraints: $9 + 3 = 12$.

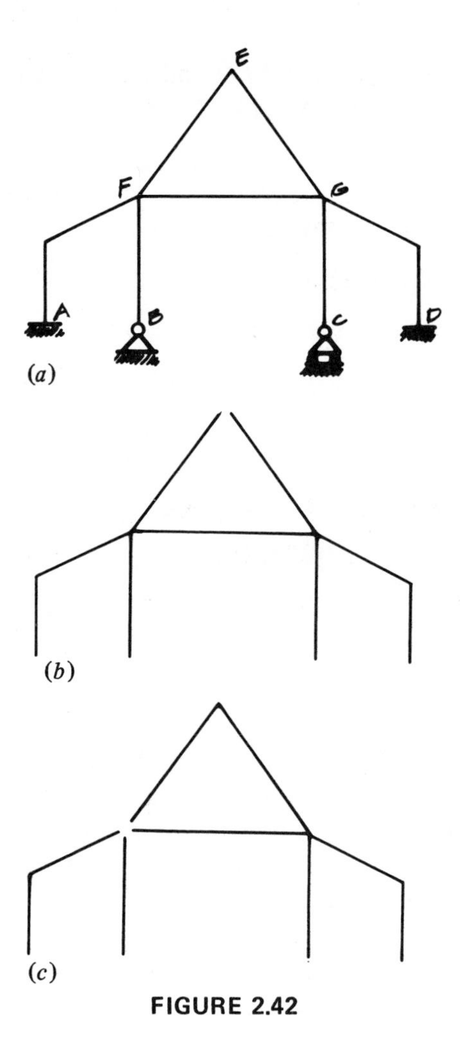

FIGURE 2.42

$$n = 1,$$

$$3n = 3 \text{ (degrees of freedom)},$$

$$12 > 3 \text{ (redundant structure)},$$

$$12 - 3 = 9 \text{ (redundant constraints)}.$$

Solution 2 (Figure 2.42*c*).

$$3f + 2h + s = 9 \text{ (subtotal of external constraints)}.$$

Weld F forms the closed loop EFG and joins bars FE, FG, FB, and FA. Hence, $3(b_F - 1) = 3(4 - 1) = 9$. Total of external and internal constraints: $9 + 9 = 18$.

$n = 3$ (bars FA, FB, and the rest of the frame is one piece without loops),

$3n = 9$ (degrees of freedom),

$18 > 9$ (redundant structure),

$18 - 9 = 9$ (redundant constraints as in Solution 1).

2.9. (Figure 2.43).

Solution 1 (Figure 2.43*b*).

$s = 1$ (roller D),

$h = 1$ (slide C),

$f = 1$ (end A),

$3f + 2h + s = 6$ (subtotal of external constraints).

Internal hinge B connects three bar ends. Hence, $2(b_B - 1) = 2(3 - 1) = 4$. After considering and discarding external constraints and the hinged connection B, the frame is free in space but contains four closed loops. They may be opened by discarding welds E, F, G, and H, after which the frame lacks closed loops.

At weld E, $3(b_E - 1) = 3(2 - 1) = 3$.
At weld F, $3(b_F - 1) = 3(2 - 1) = 3$.
At weld G, $3(b_G - 1) = 3(2 - 1) = 3$.
At weld H, $3(b_H - 1) = 3(2 - 1) = 3$.
Subtotal of internal connections: $4 + 3 + 3 + 3 + 3 = 16$.
Total of external and internal constraints: $6 + 16 = 22$.

$n = 1$,

$3n = 3$ (degrees of freedom),

$22 > 3$ (redundant structure),

$22 - 3 = 19$ (redundant constraints).

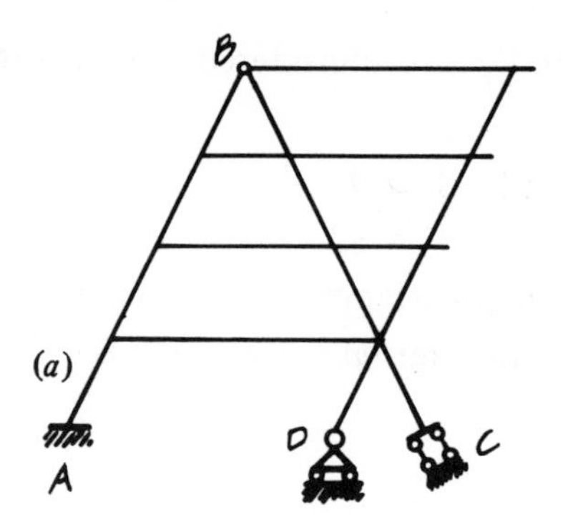

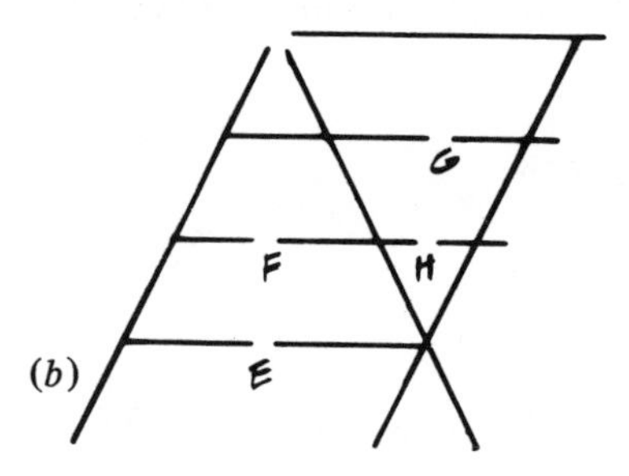

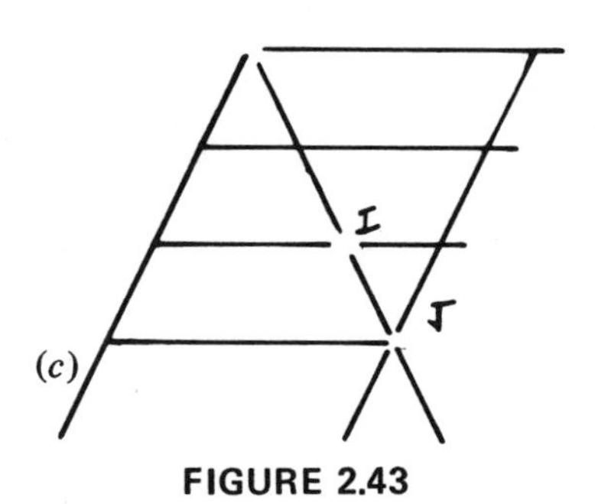

FIGURE 2.43

Solution 2 (Figure 2.43*c*).

$$3f + 2h + s = 6,$$

$$2(b_B - 1) = 4.$$

Welds at joints I and J are considered and discarded to eliminate closed loops.

At I, $3(b_I - 1) = 3(4 - 1) = 9$;
at J, $3(b_J - 1) = 3(5 - 1) = 12$.
Subtotal of internal constraints: $4 + 9 + 12 = 25$.
Total of external and internal constraints: $6 + 25 = 31$.

$$n = 4,$$

$$3n = 12 \text{ (degrees of freedom)},$$

$$31 > 12 \text{ (redundant structure)},$$

$$31 - 12 = 19 \text{ (redundant constraints as in Solution 1)}.$$

2.10. (Figure 2.44*a*).

Solution (Figure 2.44*b*).

$$s = 0,$$

$$h = 0,$$

$$f = 2 \text{ (fixed ends } A \text{ and } B),$$

$$3f + 2h + s = 6 \text{ (subtotal of external constraints)}.$$

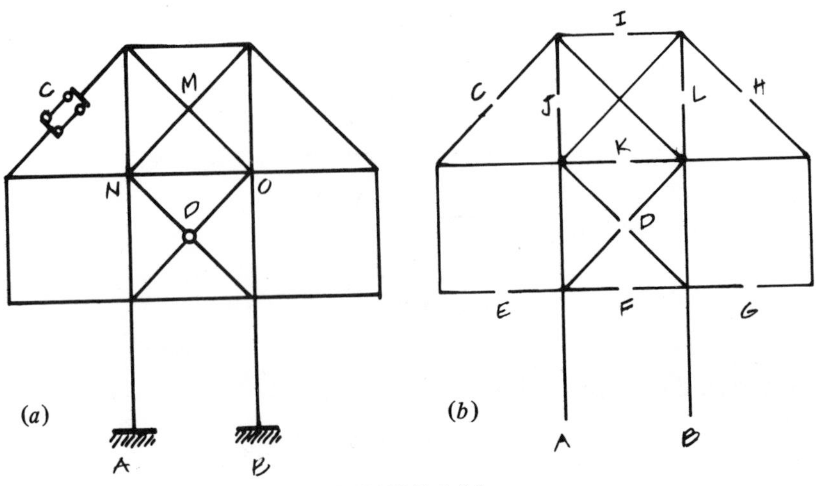

FIGURE 2.44

Internal slide C connects two bar ends. Hence, $2(b_C - 1) = 2$. Internal hinge D connects four bar ends. Hence, $2(b_D - 1) = 6$. After considering and discarding internal connections C and D, the frame has eight closed loops. These may be considered formed by welds at points E, F, G, H, I, J, K, and L, each of which connects two bar ends. Hence, $8[3(2 - 1)] = 24$

Subtotal of internal constraints: $2 + 6 + 24 = 32$.

Total of external and internal constraints: $6 + 32 = 38$.

$$n = 1,$$

$$3n = 3 \text{ (degrees of freedom)},$$

$$38 > 3 \text{ (redundant structure)},$$

$$38 - 3 = 35 \text{ (redundant constraints)}.$$

2.11. Consider internal welds at points M, N, and O of the frame rather than at points E, F, G, H, I, J, K, and L (Figure 2.44*a*). Solve and find the same answer as in Problem 2.10.

2.12. (Figure 2.45*a*).

Solution (Figure 2.45*b*).

$$s = 0,$$

$$h = 3 \text{ (connections } A, B, \text{ and } D),$$

$$f = 1 \text{ (end } C),$$

$$3f + 2h + s = 3 + 6 = 9 \text{ (subtotal of external constraints)}.$$

Internal hinged connection E joins two pieces. Hence, $2(b_E - 1) = 2$. The same can be said for connections F, G, and H. After considering and discarding connections A, B, C, D, E, F, G, and H, the frame has three closed loops, which are formed by welds at I, J, and K. Each weld joins two ends. Hence, $3[3(2 - 1)] = 9$.

Subtotal of internal constraints: $8 + 9 = 17$.

Total of external and internal constraints: $9 + 17 = 26$.

$$n = 2,$$

$$3n = 6 \text{ (degrees of freedom)},$$

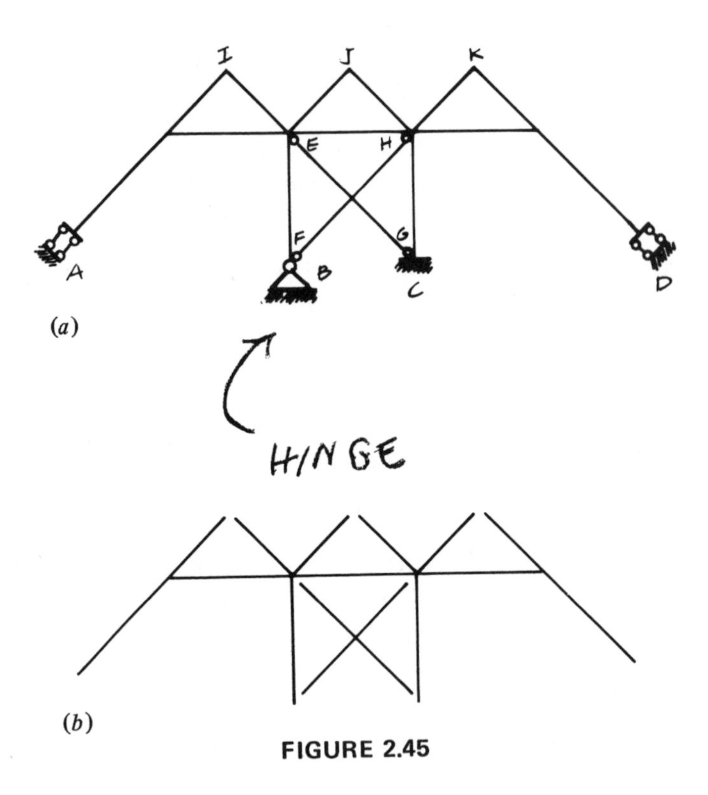

FIGURE 2.45

$$26 > 6 \text{ (redundant structure)},$$

$$26 - 6 = 20 \text{ (redundant constraints)}.$$

2.13. (Figure 2.46); solve.

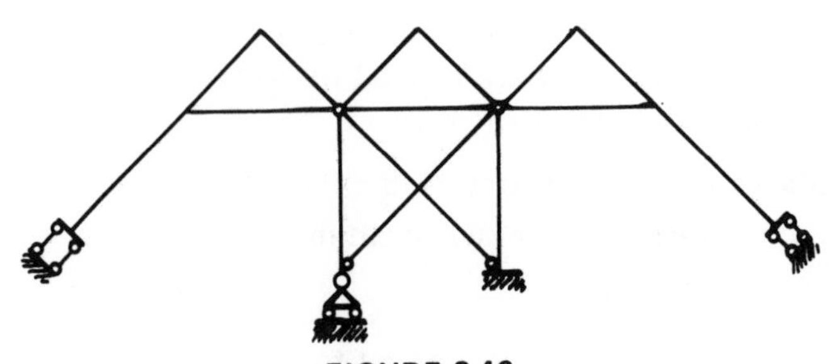

FIGURE 2.46

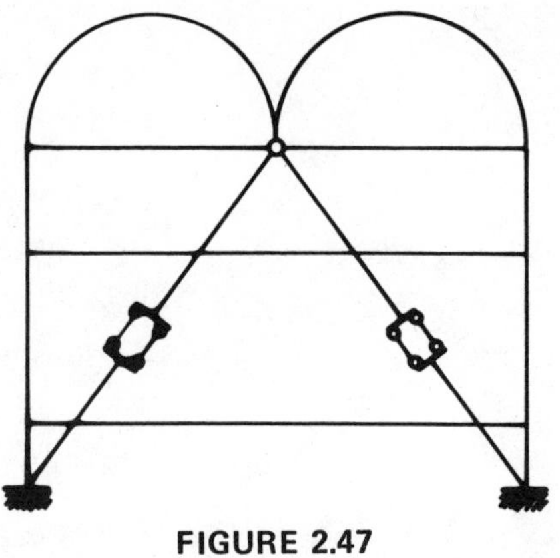

FIGURE 2.47

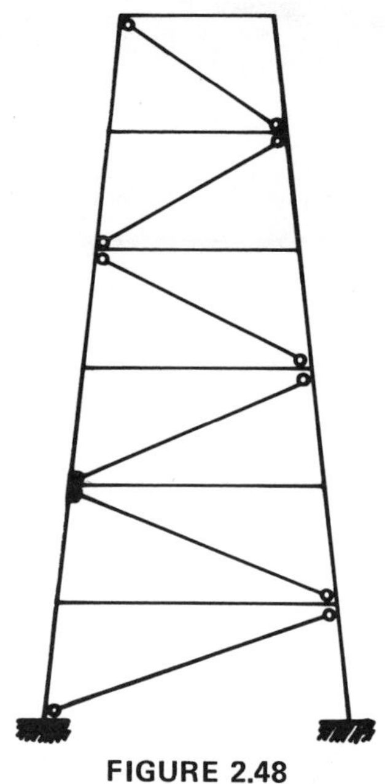

FIGURE 2.48

2.14. (Figure 2.47); solve.

2.15. (Figure 2.48); solve.

View of the interior of the Automobile Exhibit Hall in Turin, Italy. Structures by Riccardo Morandi, Rome. Photograph courtesy of Professor Morandi.

THREE

EXTERNAL FORCES: REACTIONS OF THE CONSTRAINTS

Defining the model structure is the creative part of structural design. Guided by previous experience, zoning laws, building codes, and budgetary limitations, architects and engineers, individually or in teams, select the most appropriate structure for the activities for which a building is commissioned. The shape and size of the structure and its components are chosen (geometric data) as well as the types of foundation and connections for the various structural elements (constraint data). In addition, the loads that the structure must carry in relation to its occupancy, size, and shape are all assessed (loading data).

At this point, the design takes on an analytic character. Calculations must be performed with the information supplied by the structural scheme to assure that sizes and shapes assigned to the whole structure and its parts are adequate to permit safe, comfortable, and problem-free occupancy. This is done by comparing actual stresses and deformations with their allowable limits. Structural analysis must, therefore, provide the diagrams of external and internal forces on various elements and then determine internal stresses. Some external forces, the loads, are already evident from the structural scheme; others, the reactions of the constraints, have to be calculated.

If the structure is stable and plane, each of its parts has three degrees of freedom (x and y translation plus rotation) and three constraints on its plane. The unknown constraining reactions are then found from three *equations of equilibrium*. These equations set the sum of all external forces (loads and constraining reactions) in the x and y directions and the sum of all external moments equal to

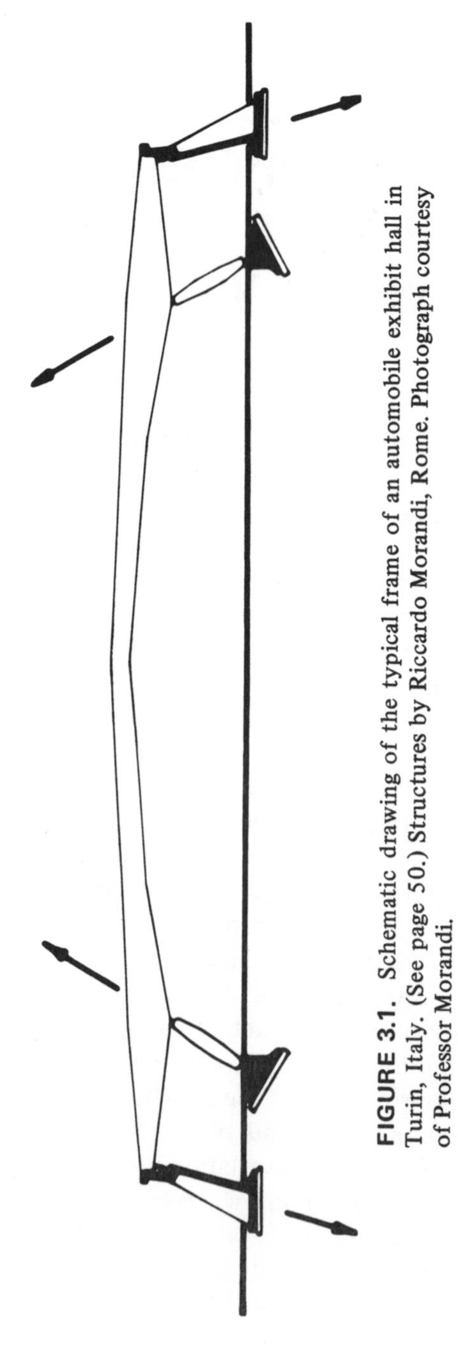

FIGURE 3.1. Schematic drawing of the typical frame of an automobile exhibit hall in Turin, Italy. (See page 50.) Structures by Riccardo Morandi, Rome. Photograph courtesy of Professor Morandi.

ca

su of external forces, t

kno the constraints w

Fed model of the st

body free from const

forces, of which

uated.

ural scheme a

of equilibrium

The first e

rigidly in the

must be zero. I

ture i

hori

yields the numeri

the second equatic

forces must be zero,

solution

provides the value of a

the third equation sets

base of the column equ

around that point (or ar

$$M = -g\frac{h^2}{2} - P$$

provides the numerical v

the reactive moment of th

Because the reactions

equations in statics, these

minate. Overconstrained s

THREE

EXTERNAL FORCES: REACTIONS OF THE CONSTRAINTS

Defining the model structure is the creative part of structural design. Guided by previous experience, zoning laws, building codes, and budgetary limitations, architects and engineers, individually or in teams, select the most appropriate structure for the activities for which a building is commissioned. The shape and size of the structure and its components are chosen (geometric data) as well as the types of foundation and connections for the various structural elements (constraint data). In addition, the loads that the structure must carry in relation to its occupancy, size, and shape are all assessed (loading data).

At this point, the design takes on an analytic character. Calculations must be performed with the information supplied by the structural scheme to assure that sizes and shapes assigned to the whole structure and its parts are adequate to permit safe, comfortable, and problem-free occupancy. This is done by comparing actual stresses and deformations with their allowable limits. Structural analysis must, therefore, provide the diagrams of external and internal forces on various elements and then determine internal stresses. Some external forces, the loads, are already evident from the structural scheme; others, the reactions of the constraints, have to be calculated.

If the structure is stable and plane, each of its parts has three degrees of freedom (x and y translation plus rotation) and three constraints on its plane. The unknown constraining reactions are then found from three *equations of equilibrium*. These equations set the sum of all external forces (loads and constraining reactions) in the x and y directions and the sum of all external moments equal to

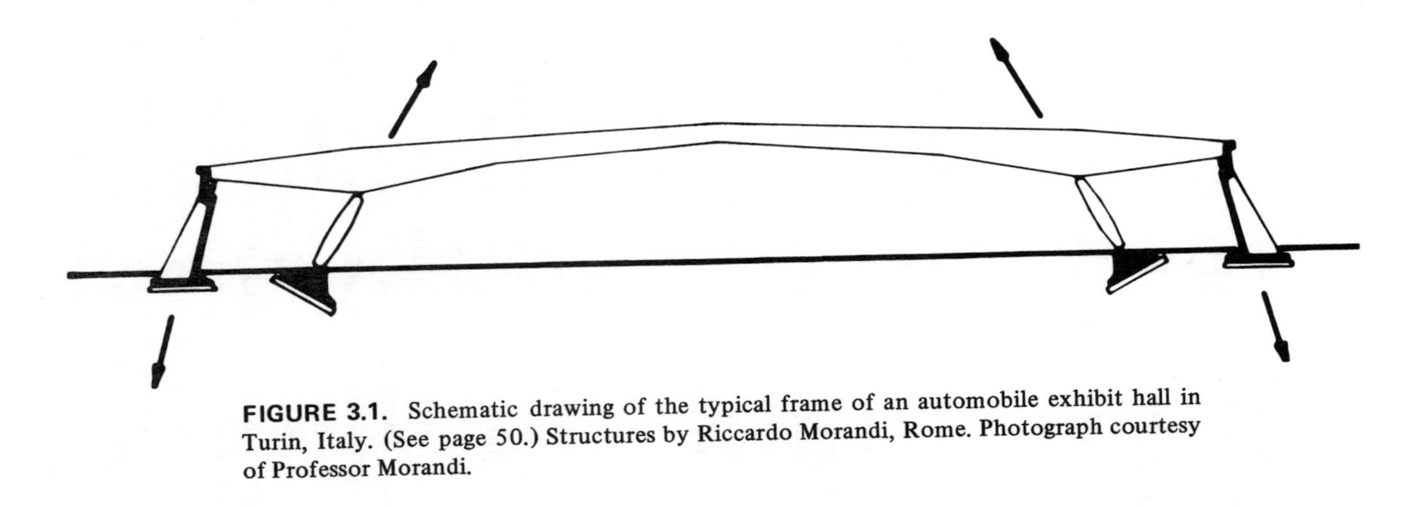

FIGURE 3.1. Schematic drawing of the typical frame of an automobile exhibit hall in Turin, Italy. (See page 50.) Structures by Riccardo Morandi, Rome. Photograph courtesy of Professor Morandi.

zero. In order to have a clearer diagram of external forces, the structural scheme is modified by replacing the constraints with their reactive forces and moments. This modified model of the structure is called a *free body diagram*, because it is free from constraints but subject to a balanced system of external forces, of which some are known (the loads) and some have to be evaluated.

Figures 3.2*a* and *b* show a plane structural scheme and its free body diagram, for which the three equations of equilibrium are:

$$\Sigma F_x = 0 = gh - H,$$

$$\Sigma F_y = 0 = P - V,$$

$$\Sigma M = 0 = -M - (gh)\frac{h}{2} - P\frac{l}{2}.$$

The first equation indicates that since the structure is not moving rigidly in the horizontal direction, the sum of all horizontal forces must be zero. Its solution

$$H = gh = 0.5(10) = 5 \text{ k}$$

yields the numerical value 5 k of the unknown reaction H. Similarly, the second equation indicates that the sum of all external vertical forces must be zero, since the structure does not move vertically. Its solution

$$V = P = 2 \text{ k}$$

provides the value of another external force, the reaction V. Finally, the third equation sets the sum of all external moments around the base of the column equal to zero, since the structure does not rotate around that point (or any other). Its solution

$$M = -g\frac{h^2}{2} - P\frac{l}{2} = -0.5\frac{(10)^2}{2} - 2\frac{(12)}{2} = -37 \text{ k-ft.}$$

provides the numerical value of the last generalized external force, the reactive moment of the base.

Because the reactions of stable structures are found by solving equations in statics, these structures are also called *statically determinate*. Overconstrained structures are called *statically indeterminate*

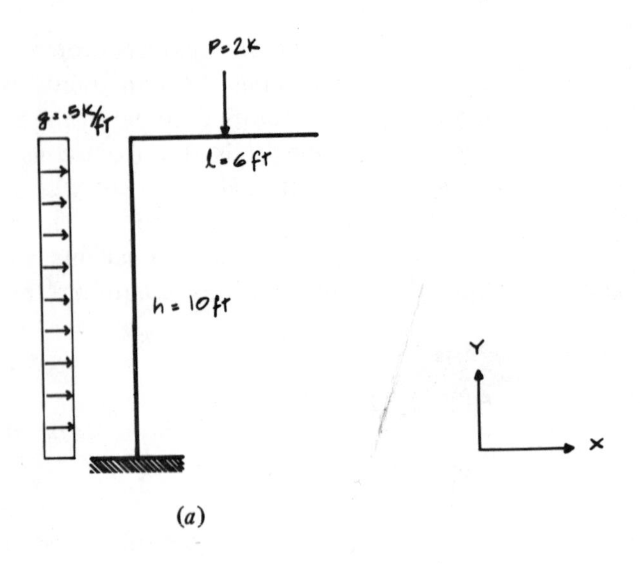

(a)

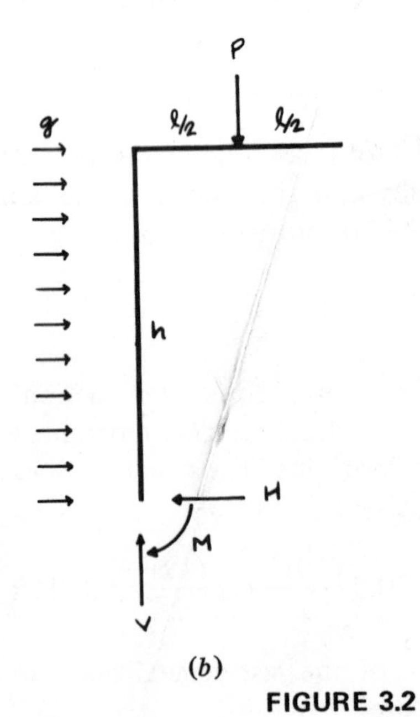

(b)

FIGURE 3.2

by H_r/A. Defining the strain as the ratio $\Delta L/L$ and the Young's modulus as the ratio of stress to strain, we have

$$E = \left(\frac{H_r}{A}\right) \div \left(\frac{\Delta L}{L}\right);$$

hence,

$$\Delta L = \frac{H_r L}{EA}.$$

The thermal expansion and the shrinkage due to the unknown load H_r, which is the reaction of the redundant constraint, are now algebraically summed and set equal to the deformation of the original structure, which is zero, since the two terminal hinges do not permit either expansion or contraction.

$$\alpha L \Delta T - \frac{LH_r}{EA} = 0,$$

from which

$$H_r = EA\alpha\Delta T.$$

For a steel bar,

$$\alpha = \frac{65}{(10)^7} \frac{1}{°\text{F}}.$$

$$E = 30{,}000 \text{ k/in}^2.$$

If

$$A = 2 \text{ in.}^2$$

and

$$\Delta T = 50 \text{ °F}$$

then

$$H_r = \frac{30{,}000\ (2)\ 50\ (65)}{(10)^7} = 19.5 \text{ k}.$$

The statically determinate reaction H_l is found from the equilibrium of the free body (Figure 3.3*b*).

$$\Sigma F_x = 0 = H_l - H_r; \quad H_l = H_r = 19.5 \text{ k}.$$

H_l, of course, could be considered the redundant reaction and H_r the statically determinate one. The choice of the reactions to be evaluated by equilibrium and those to be evaluated otherwise does not affect the value of any of the reactions.

In forthcoming chapters, dedicated to the analysis of reactions and internal forces for various types of structural elements, only stable structures will be considered unless otherwise stated.

PROBLEMS

Evaluate the reactions of constraints on the following structures.

3.1. (Figures 3.4*a* and *b*).

Solution. The free body diagram of the portal in Figure 3.4*a* is shown in Figure 3.4*b*, where unknown reactions V_L and H_L replace the external hinge; reaction V_R replaces the link (right column); and the distributed wind load is replaced by its resultant. The equation of horizintal equilibrium is

$$\Sigma F_x = 0 = 180 + H_L,$$

from which

$$H_L = -180 \text{ lb}.$$

According to the minus sign in the result, H_L has the sign opposite to that assumed in the free body diagram. The equation of moment equilibrium around the hinge is

$$\Sigma M = 0 = -180(6) - 1000(6) + 12\, V_R,$$

which yields

$$V_R = 180\,\frac{(6)}{12} + 1000\,\frac{(6)}{12} = 590 \text{ lb}.$$

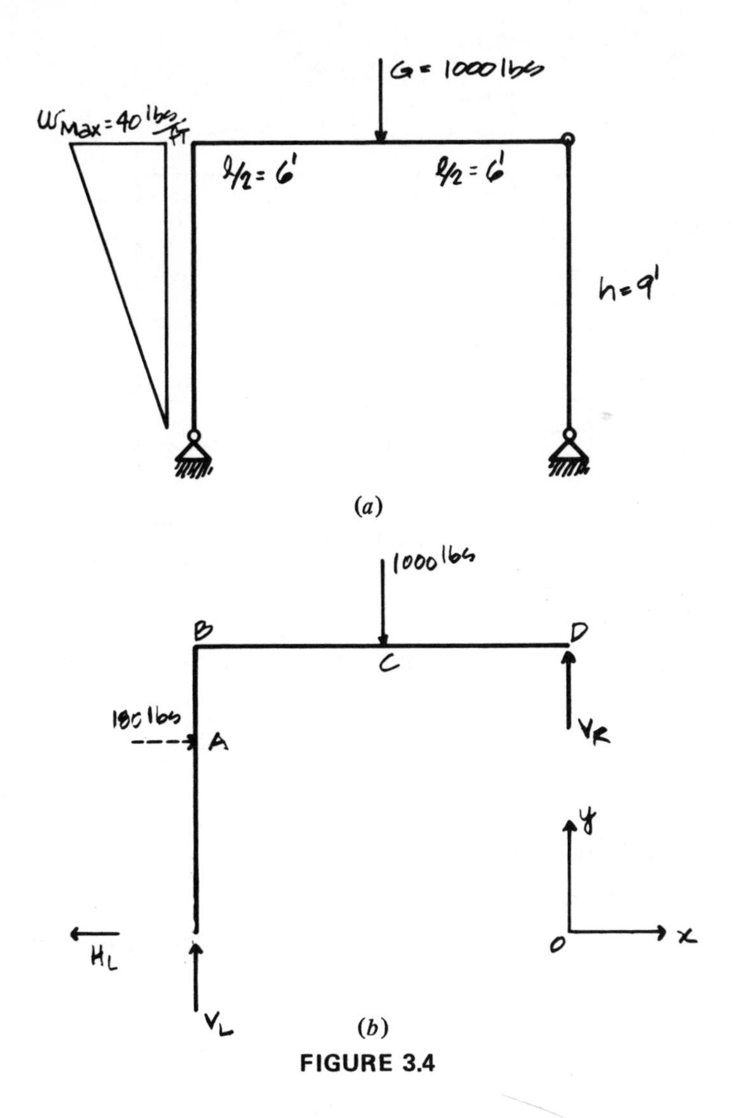

FIGURE 3.4

The equation of vertical equilibrium is

$$\Sigma F_y = 0 = V_L - 1000 + 590,$$

from which

$$V_L = 410 \text{ lb.}$$

3.2. (Figure 3.5); solve for reactions H_L, V_L, and V_R.

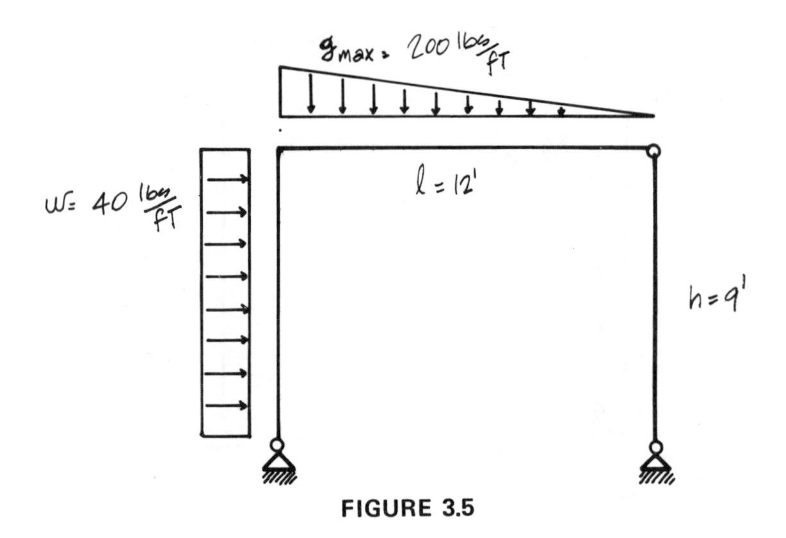

FIGURE 3.5

3.3. (Figures 3.6*a* and *b*).

Solution. The equation of horizontal equilibrium is

$$\Sigma F_x = 0 = 2 + 2 - H_R\,;$$

thus,

$$H_R = 4 \text{ k}.$$

The equation of moment equilibrium around the hinge is

$$-120 V_L + 10(100) - 2(15) + 10(80) + 10(60) + 10(40) - 2(15) + 10(20) = 0,$$

from which

$$V_L = \frac{1}{120}(1000 - 30 + 800 + 600 + 400 - 30 + 200)$$

$$= \frac{2940}{120} = 24.5 \text{ k}$$

The equation of vertical equilibrium is

$$\Sigma F_y = 0 = 24.5 - 5(10) + V_R\,;$$

thus,

$$V_R = 25.5 \text{ k}.$$

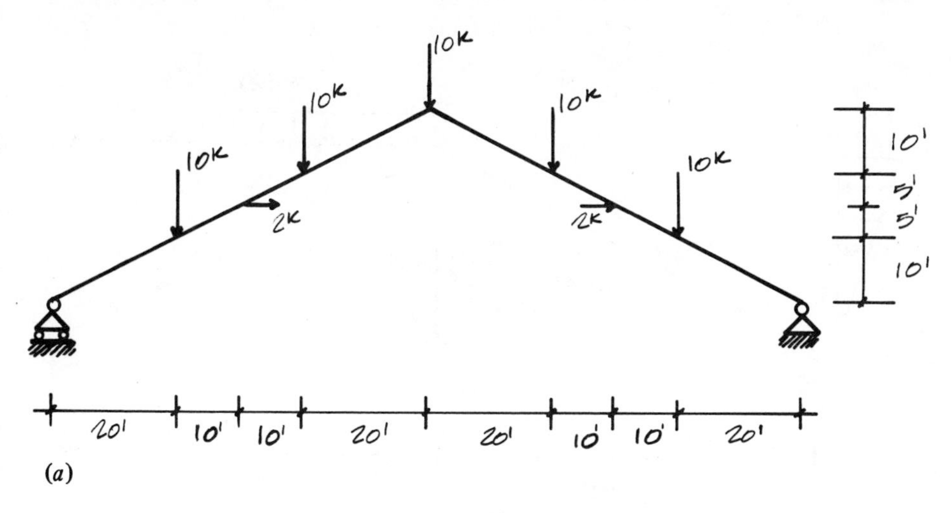

(*a*)

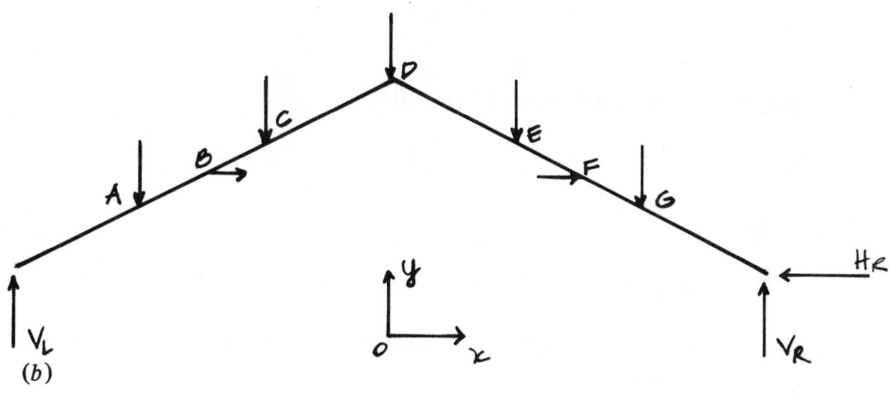

(*b*)

FIGURE 3.6

3.4. (Figures 3.8*a* and *b*).

Solution.

$$\Sigma F_x = 0 = 10 + 20 + 10 - R_R ,$$

thus,

$$R_R = 40 \text{ k}.$$

$$\Sigma F_y = 0 = R_L - 10 - 20 + 10,$$

from which

$$R_L = 20 \text{ k}.$$

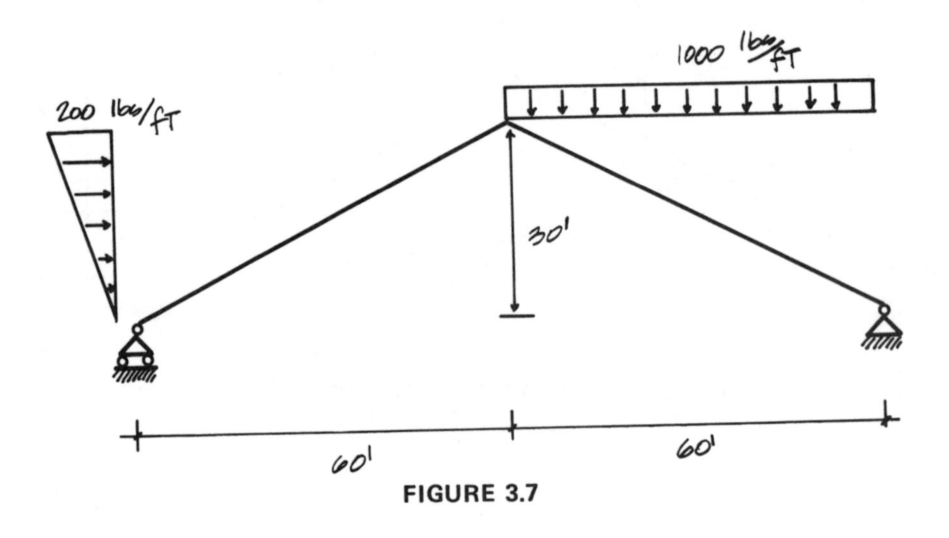

FIGURE 3.7

Equilibrium of moments about the roller

$$\Sigma M = 0 = M_L - 80(20) \cos 45° + 14.2(60) + 28.4(40) - 14.2(10),$$

from which

$$M_L = 14.2(10 - 80 - 60 + 80) = -710 \text{ k}.$$

The sign of M_L is the opposite of the assumed sign, since the solution of the equation is negative.

3.5. (Figure 3.7); solve for V_L, V_R, and H_R.

3.6. (Figure 3.9); solve for R_L, R_R, and M_L.

3.7. (Figures 3.10*a* and *b*).

Solution. Inspecting the frame geometries,

$$R_R = \frac{H_R}{\cos 60°} = 2H_R,$$

$$V_R = H_R \tan 60° = H_R (3)^{1/2}$$

Horizontal equilibrium:

$$\Sigma F_x = 0 = -10 + H_R;$$

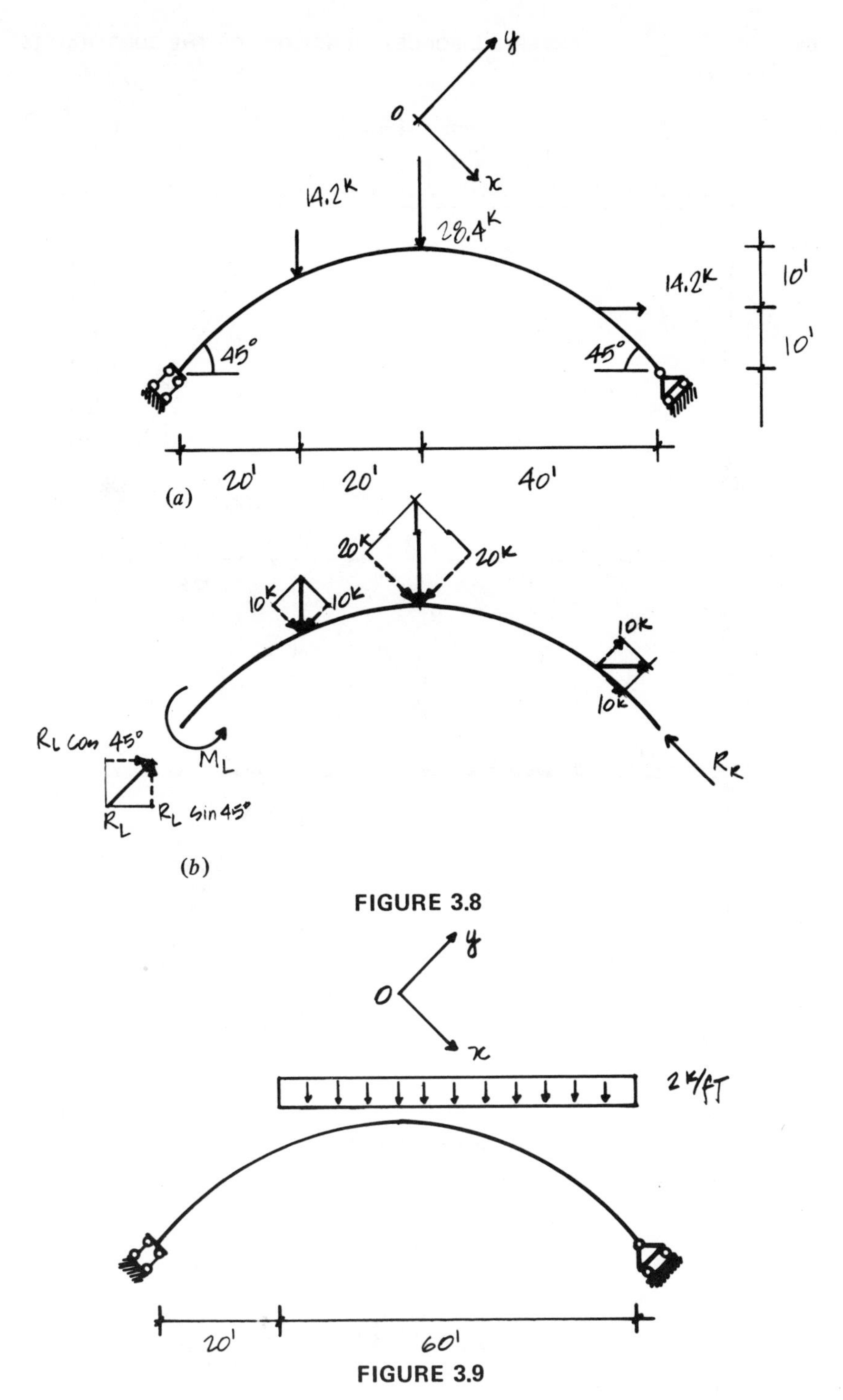

FIGURE 3.8

FIGURE 3.9

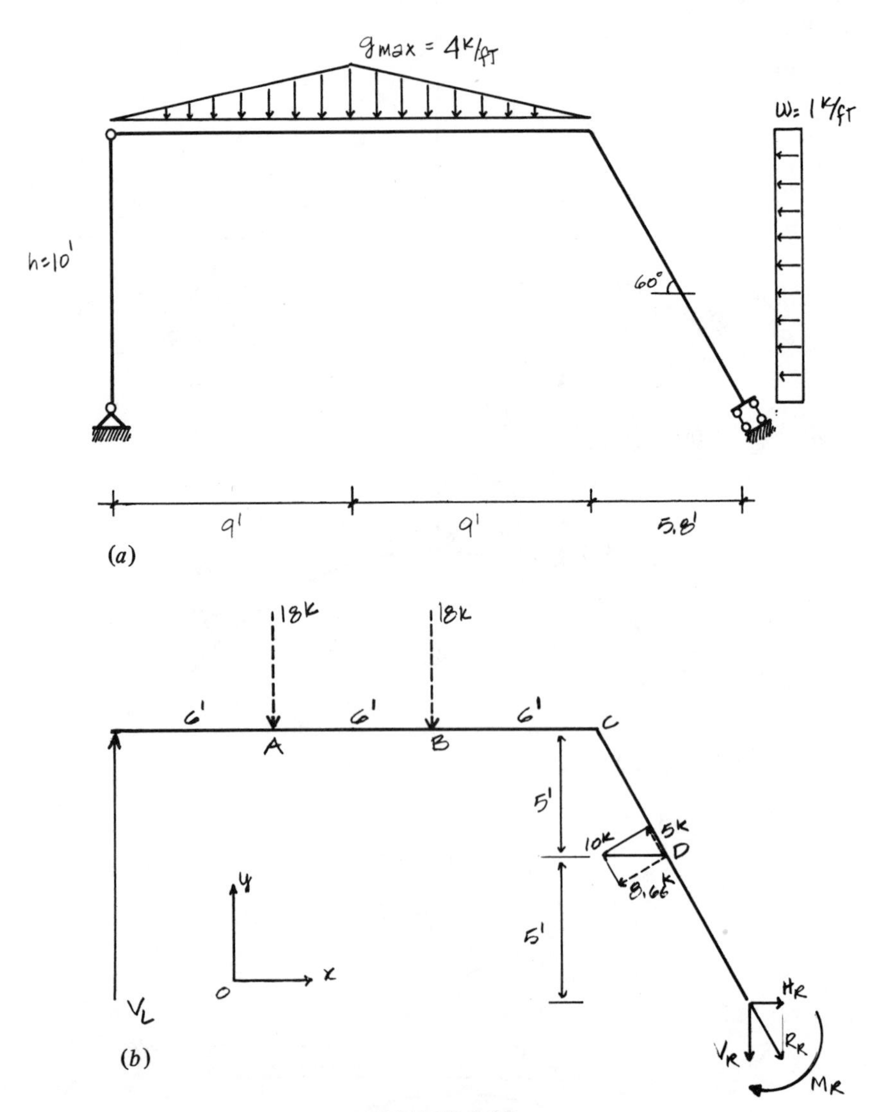

FIGURE 3.10

thus,

$$H_R = 10 \text{ k},$$

$$R_R = 20 \text{ k},$$

$$V_R = 17.32 \text{ k}.$$

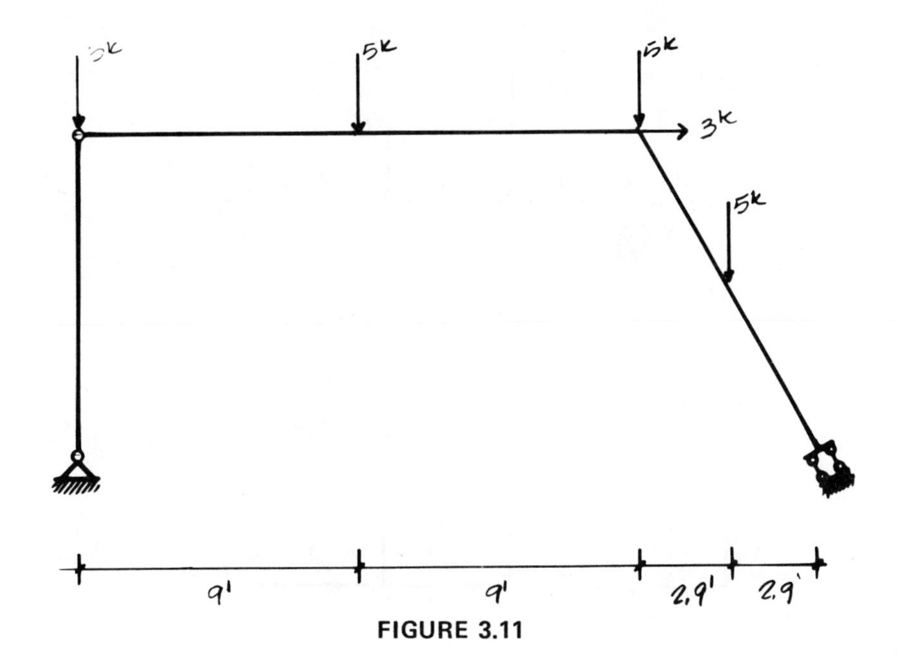

FIGURE 3.11

Vertical equilibrium:

$$\Sigma F_y = 0 = -18 - 18 + V_L - 17.32,$$

from which

$$V_L = 53.32 \text{ k}.$$

Equilibrium of moments about the foundation hinge:

$$\Sigma M = 0 = -M_R - 23.8(17.32) + 10(5) - 18(12) - 18(6);$$

therefore,

$$M_R = -412 + 50 - 216 - 108 = -686 \text{ k-ft}.$$

The assumed sign for M_R must be reversed due to the negative solution of the equilibrium equation.

3.8. (Figure 3.11); solve for V_L, V_R, H_R, and M_R.

3.9. (Figures 3.12*a* and *b*).

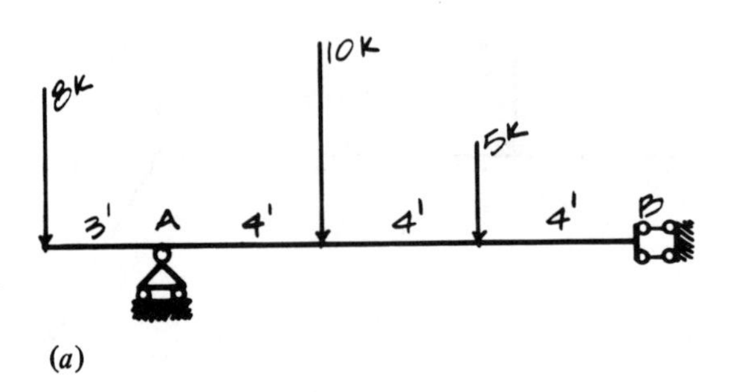

(a)

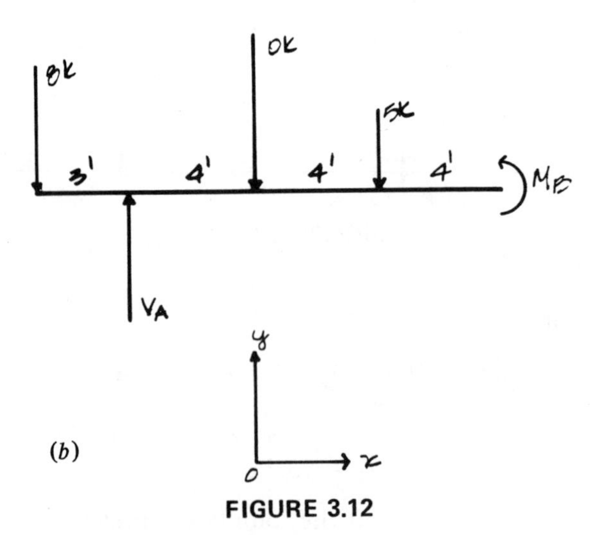

(b)

FIGURE 3.12

Solution. Vertical equilibrium:

$$\Sigma F_y = 0 = -8 - 10 - 5 + V_A;$$

thus,

$$V_A = 23 \text{ k}.$$

Equilibrium of moments about roller A:

$$\Sigma M = 0 = 8(3) - 10(4) - 5(8) + M_B,$$

from which

$$M_B = -24 + 40 + 40 = 56 \text{ k-ft}.$$

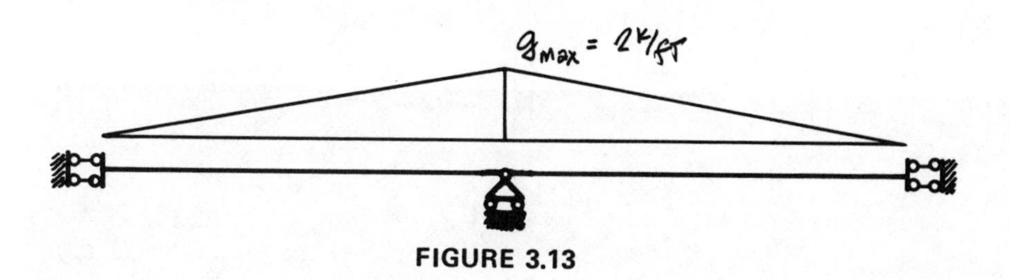

FIGURE 3.13

3.10. (Figure 3.13); solve for V and M. Note that due to symmetry,

$$M_L = M_R = M.$$

A reinforced concrete beam tested for failure by three point loads. The three marks near the top surface show the position of the loads. The distribution of the failure cracks shows that internal forces exist everywhere in the beam. Photograph courtesy of Portland Cement Association.

FOUR

INTERNAL FORCES

Loads and reactions of constraints are external forces, and they are applied at single points or on single segments of the geometric axis of structures. Internal forces, on the other hand, exist everywhere in the structural material and are not visible in the free body diagram of structural elements. Their evaluation is another step toward calculating internal stresses.

The existence of internal forces and moments can be easily demonstrated by considering, for example, a cantilever beam loaded at its tip by a vertical force F (Figure 4.1*a*). The same beam is shown in Figure 4.1*b* with a detached piece of length d at its right end. When the two pieces are connected, the left piece must support the right one with a force equal and opposite to F. Otherwise, the right piece would fall down. The right piece must discharge its load F on the left piece, or else this piece would not bend, which contradicts experimental evidence.

The equal, opposite, and parallel forces F produce a clockwise moment Fd on the right part. Hence, this part would spin without the counterclockwise moment Fd transferred to it by the left part. The clockwise moment Fd is also transferred by the right part to the left according to the principle of action and reaction. The length d of the right piece can range from a fraction of an inch to the full length of the beam. It is evident, then, that internal forces F and moments Fd exist everywhere in the structure even though they are not visible in the model in Figure 4.1*a*.

Internal forces perpendicular to the structural axis, such as the forces F in Figure 4.1*b*, are called *shear forces*. Internal forces parallel to the structural axis are called *axial forces*. Internal moments produced by forces and lever arms on the plane of the structure, such as the moments Fd in Figure 4.1*b*, are called *bending moments*. In-

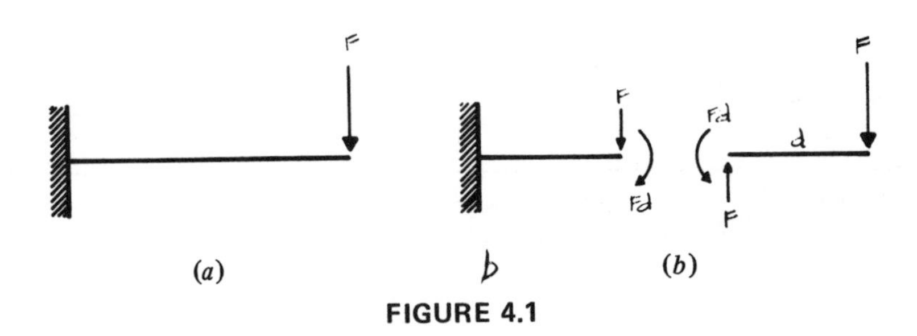

FIGURE 4.1

ternal moments produced by forces and lever arms on planes perpendicular to the structural axis are called *twisting moments*.

It is possible to visualize these four actions and their effects by representing a section of the structure with an accordion. Holding the accordion between our hands, we can push it in or pull it out, keeping the palms rigid and perfectly parallel. The accordion is subject, in this case, to compressive or tensile axial forces. The top and bottom lines of its side elevation shrink or expand equally, and the side elevation remains rectangular (Figure 4.2*a*).

Holding our palms rigidly parallel, we can move one end of the accordion up and one down; this time, the accordion is subject to shear forces. The side elevation is no longer rectangular; it looks rather like a rhombus. The horizontal sides do not change length appreciably, but one of the diagonals of the side elevation becomes longer and one becomes shorter (Figure 4.2*b*).

If, starting from a parallel position, we rotate the palm of our right hand clockwise and the left palm counterclockwise, the lower part of the accordion is pushed in, the upper part is pulled out, and the accordion is subject to bending. The upper side of the elevation stretches as it curves and the lower side shrinks. The left and right sides of the elevation remain straight and perpendicular to the curved sides (Figure 4.2*c*).

Finally, if we keep our palms rigidly parallel but point the fingertips of the right hand away and those of the left hand toward us, the accordion is subject to a twisting moment. The side elevation becomes a warped three-dimensional surface (Figure 4.2*d*). Its right and left sides rotate with the palms of our hands but remain straight. Both top and bottom sides of the elevation form a helix. One diag-

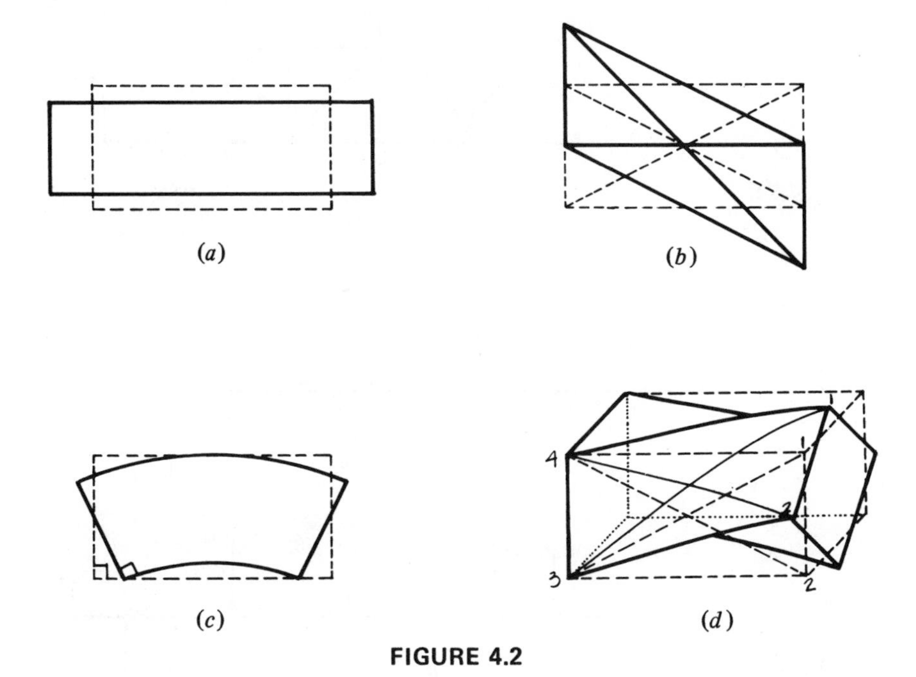

FIGURE 4.2

onal of the side elevation (2, 4) shrinks, and the other (3, 1) extends. For simplicity, in Figure 4.2*d*, only the right end of the accordion is rotated.

When all external forces and moments are known, evaluating internal forces is always a statically determinate problem even if some reactions of the constraints are redundant and calculated by methods other than equilibrium equations. It could be said that evaluating internal forces is nothing more than algebraically summing external forces or moments of the same kind.

A simple example is sufficient to justify and explain the preceding statements. We consider the free body diagram of a beam in Figure 4.3*a*, which is in Figure 4.3*b* and a cross section S 12 ft from the beam's left end. Evaluating internal forces on this or any other section proceeds as follows. A free body diagram is drawn for the beam part on the left or the right (Figure 4.3*c*) of section S. The left part of the beam is treated as a constraint on the right part and vice versa. In free body diagrams, therefore, internal forces on section S appear as unknown reactions of the constraining parts and are evaluated by

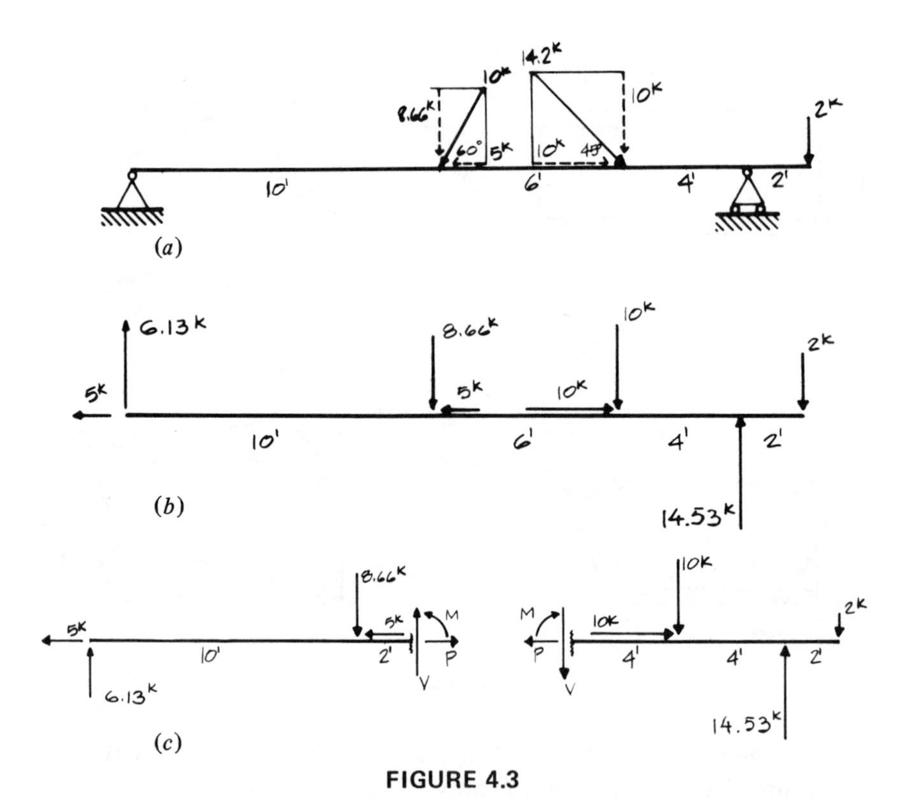

FIGURE 4.3

the three equations of plane equilibrium. Therefore, the internal forces are statically determinate. In this case, the equations of equilibrium for the left part are

$$\Sigma F_x = 0 = -5 - 5 + P,$$

$$\Sigma F_y = 0 = 6.13 - 8.66 + V,$$

$$\Sigma M_s = 0 = -12(6.13) + 2(8.66) + M.$$

The same equations can be written as

$$P = 5 + 5,$$

$$V = -6.13 + 8.66,$$

$$M = 12(6.13) - 2(8.66),$$

which shows that

internal forces on section S *are obtained by algebraically summing all forces of the same kind from the left end to* S.

Considering the right part of the beam, the equilibrium equations are

$$\Sigma F_x = 0 = -P + 10,$$

$$\Sigma F_y = 0 = -V - 10 + 14.53 - 2,$$

$$\Sigma M_x = 0 = -2(10) + 8(14.53) - 10(4) - M.$$

By writing these equations in the form

$$P = 10,$$

$$V = -2 + 14.53 - 10,$$

$$M = -2(10) + 8(14.53) - 4(10),$$

it is readily seen that

internal forces on section S *are also found by algebraically summing all forces of the same kind from the right end of the beam to section* S.

In the chapters dedicated to beams, cables, trusses, and arches, we increase our understanding of internal forces by concentrating on their evaluation, which is the third step in the systematic approach to structural analysis.

It was mentioned earlier in Chapter 4 that internal forces exist everywhere in the structural material. It is therefore important for a true understanding of structural behavior and design to learn the distribution of internal forces along the axis of structural elements. Moreover, awareness of internal force distribution can be used to save on construction materials and achieve structural elegance with geometric design reflecting the flow of internal forces from loads to supports.

The four kinds of internal forces are not usually found together in plane structural elements. Rather, depending on the geometries, loads, constraints, and rigidity of each element, only one or a few internal forces prevail under theoretically stated conditions.

Flexural members, such as horizontal beams under vertical loads, are subject to only shear forces and bending moments; tension ele-

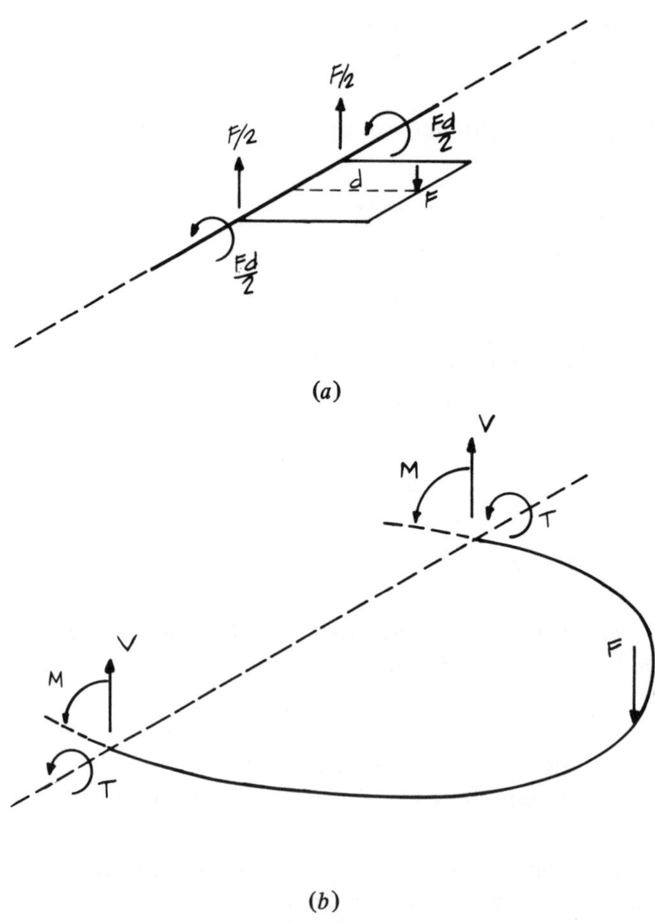

FIGURE 4.4

ments, such as hangers, cables, and the tension bars of trusses, are subject to only axial forces. Compression members, like columns and compression bars of trusses, are also subject to only axial forces. Structural elements, such as columns loaded vertically and horizontally and arches, carry bending moments as well as axial and shear forces. Twisting moments coexist with shear forces and bending moments in such structural components as spandrel beams carrying balcony slabs and beams curved on the horizontal plane (Figures 4.4*a* and *b*).

The separate distributions of these internal forces and moments along the geometric axis of a structural element are called the *shear diagram*, *axial force diagram*, *bending moment diagram*, and *twisting moment diagram*. These four types of diagrams can be drawn in appropriate scale or even sketched freehand with numerical indications of relevant values of internal forces and also the degree of the geometric curve by which they are represented; for example, $1°$ for a linear diagram $2°$ for a parabolic diagram, and so forth.

In the following chapters, we consider diagrams of internal forces for such structural elements as beams, cables, trusses, and arches. These structures are characterized precisely by the presence of some of the four kinds of internal forces and by the absence of others.

PROBLEMS

4.1. Draw a diagram of axial forces on the frame (Figure 3.4*b*).

Solution. The axial force at any section of the column is the sum of all vertical external forces from the hinge to the section being considered. From the hinge to point B, the only external vertical force to be considered is V_L. Hence, the axial force in the column is equal everywhere to V_L (Figure 4.5*a*).

The axial force at any section of beam BD is the sum of all horizontal external forces from the end D (or B) to the section being considered. Due to the lack of horizontal external forces from D to B, the axial force is zero everywhere on the beam (Figure 4.5*a*).

4.2. Draw a diagram of shear forces on the frame (Figure 3.4*b*).

Solution. The shear force at any section of the column is the sum of all horizontal external forces from the hinge to the section being considered. At the hinge, the external horizontal force is H_L = 180 lb, with the sign of the negative x-semiaxis. Due to lack of additional horizontal external forces from the hinge to point A, the column shear keeps the constant value of 180 lb. At point A, a 180-lb horizontal external force with the sign of the positive x-semiaxis makes the shear vanish. From A to B, the shear keeps the constant value of 0 lb for lack of additional horizontal external forces (Figure 4.5*b*).

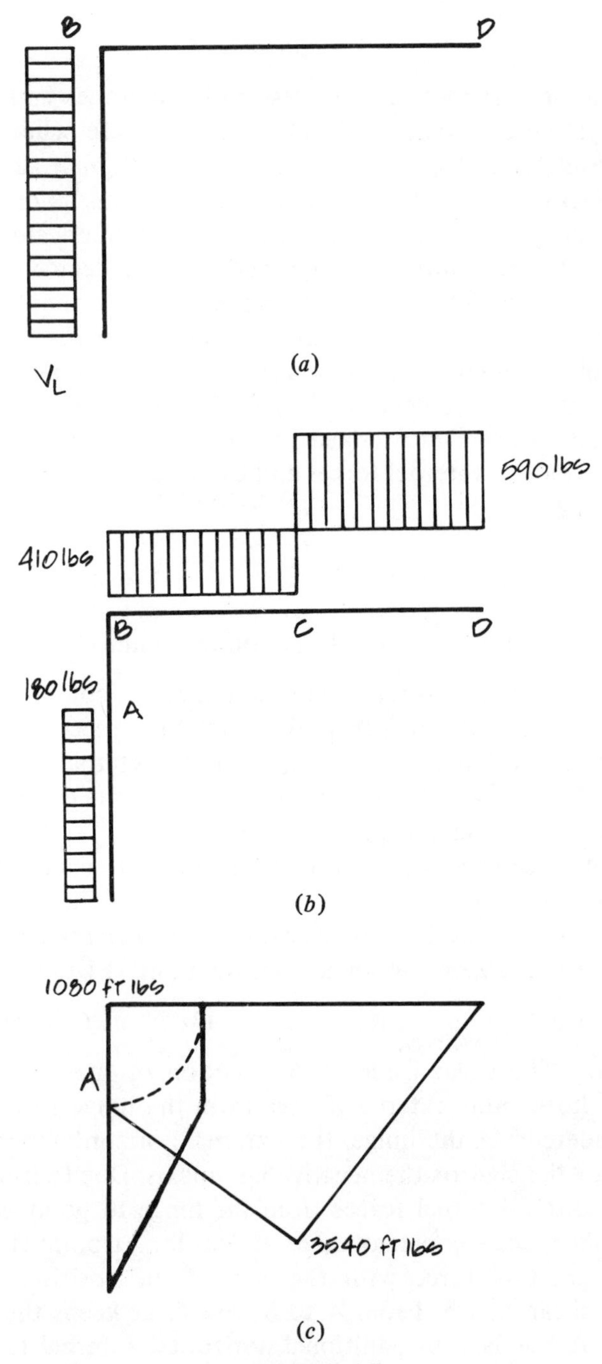

FIGURE 4.5

The shear force at any section of the beam is the sum of all vertical external forces from the end D (or B) to the section being considered. At the end D, the external vertical force is $V_R = 590$ lb, with the sign of the positive y-semiaxis. The shear keeps the constant value of 590 lb from D to C for lack of additional vertical external forces. At point C, a 1000-lb vertical load with the sign of the negative y-semiaxis changes the shear force from +590 to -410 lb. This new value remains constant from C to B for lack of additional loads. At point B, the reaction $V_L = 410$ lb, transferred through the column from the hinge to point B, makes the shear vanish (Figure 4.5*b*).

4.3. Draw a diagram of bending moments on the frame (Figure 3.4*b*).

Solution. Starting from the hinge where the moment is zero, we can readily see that reaction V_L does not produce bending moments in the column due to a lack of lever arms around the points of the column. Reaction $H_L = 180$ lb produces a clockwise moment $1(180) = -180$ ft lb at a point 1 ft away from the hinge. At points 2, 3, or 4 ft away from the hinge, H_L produces these moments: $-2(180)$ ft lb, $-3(180)$ ft lb, $-4(180)$ ft lb, and so forth, all showing a linear increase of moments along the column (Figure 4.5*c*). Accordingly, the moment at point A has the value $M_A = -6(180) = -1080$ ft lb. At a section 1 ft above point A, reaction H_L produces a clockwise moment $-7(180)$ ft lb, while the 180-lb load produces a counterclockwise moment $+1(180)$ lb. The algebraic sum is $-6(180) = -1080$ ft lb. Reasoning similarly for moments at points 2 ft and 3 ft, and so on, above point A, we see that the bending moment diagram maintains the constant value of -1080 ft lb from A to B (Figure 4.5*c*).

The moment diagram on the beam can be started more simply from end D, where V_R does not produce a moment due to the absence of a lever arm around D. At points to the left of D, reaction $V_R = 590$ lb produces counterclockwise moments that increase linearly from $M_D = 0$ to $M_C = 6(590) = 3540$ ft lb. At a point 1 ft away from the left of point C, V_R produces a counterclockwise moment 7(590) ft lb, while the 1000-lb load produces a clockwise moment $-1(1000)$ ft lb. The algebraic sum is 3130 ft lb, and it shows that the moment starts decreasing after attaining its largest value M_C.

In general, at a point x ft away from end D ($x > 6$ ft), the bending moment is

$$M(x) = xV_R - 1000(x - 6),$$

a linear function of the distance from end D. At point B, then, the bending moment on the beam is

$$M_B = 12(590) - 6(1000) = 1080 \text{ ft lb}.$$

This counterclockwise moment is indeed equal and opposite to moment M_B evaluated on the column, as required by the equilibrium of section B (Figure 4.5c).

4.4. Draw a diagram of bending moments on the beam of the frame starting from end B (Figures 3.4b and 4.5c).

Solution. The column is considered to be a constraint of the beam. Its reactions to the load on the beam are a vertical force V_L = 410 lb, a horizontal force $H_L - 180 = 0$ lb, and a clockwise moment $M_B = -1080$ ft lb (see Problem 4.3).

At a section x ft away from B, V_L produces a clockwise moment xV_L that is added to M_B. Thus, the total bending is $M_B + xV_L = -1080 - 410x$, which shows that the moment increases linearly from point B to point C, where

$$M_C = -1080 - 6(410) = -1080 - 2460 = -3540 \text{ ft lb}.$$

When x exceeds 6 ft, the counterclockwise moment produced by the 1000-lb load must be taken into account. For example, for $x = 7$ ft,

$$M = -1080 - 7(410) + (7 - 1)1000 = 2950 \text{ ft lb},$$

which shows that the moment decreases on the right side of point C. In general, for $x > 6$ ft,

$$M(x) = -1080 - x(410) + (x - 6)1000,$$

a linear function of the distance from the beam end B. For $x = 12$ ft (end D),

$$M_D = -1080 - 12(410) + 1000(6) = 0 \text{ ft lb}.$$

4.5. Draw the axial force diagram for the left part of the structure (Figure 3.6b).

Solution. Reaction $V_L = 24.5$ k, the 10-k vertical loads, and the 2-k horizontal load are first decomposed into their axial and shear components. The axis of the structure is inclined with an angle α on the horizontal plane such that

$$\cos\alpha = \frac{2}{(5)^{1/2}}, \quad \sin\alpha = \frac{1}{(5)^{1/2}}.$$

A vertical force V has, therefore, axial component A and shear components S, respectively, given by

$$A = V\sin\alpha = \frac{V}{(5)^{1/2}}, \quad S = V\cos\alpha = \frac{2V}{(5)^{1/2}}.$$

The free body diagram in Figure 3.6*b* is shown again in Figure 4.6 with the axial and shear components of the external forces.

At the roller, the axial force is

$$A_L = \frac{V_L}{(5)^{1/2}} = \frac{24.5}{(5)^{1/2}}\ \text{k}.$$

The axial force diagram remains constant from the roller to point A for lack of additional external forces (Figure 4.6). At point A, the axial component $10/(5)^{1/2}$ k of the 10-k load reduces the axial force to

$$A_A = A_L - \frac{10}{(5)^{1/2}} = \frac{14.5}{(5)^{1/2}}\ \text{k}.$$

This value remains constant as far as point B, where the axial component $4/(5)^{1/2}$ k of the 2-k horizontal load increases the axial force to

$$A_B = A_A + \frac{4}{(5)^{1/2}} = \frac{18.5}{(5)^{1/2}}\ \text{k}.$$

This value remains constant as far as point C, where the axial component $10/(5)^{1/2}$ k of the local vertical load reduces the value of the axial force to

$$A_C = A_B - \frac{10}{(5)^{1/2}} = \frac{8.5}{(5)^{1/2}}\ \text{k}.$$

This axial force remains constant as far as point D.

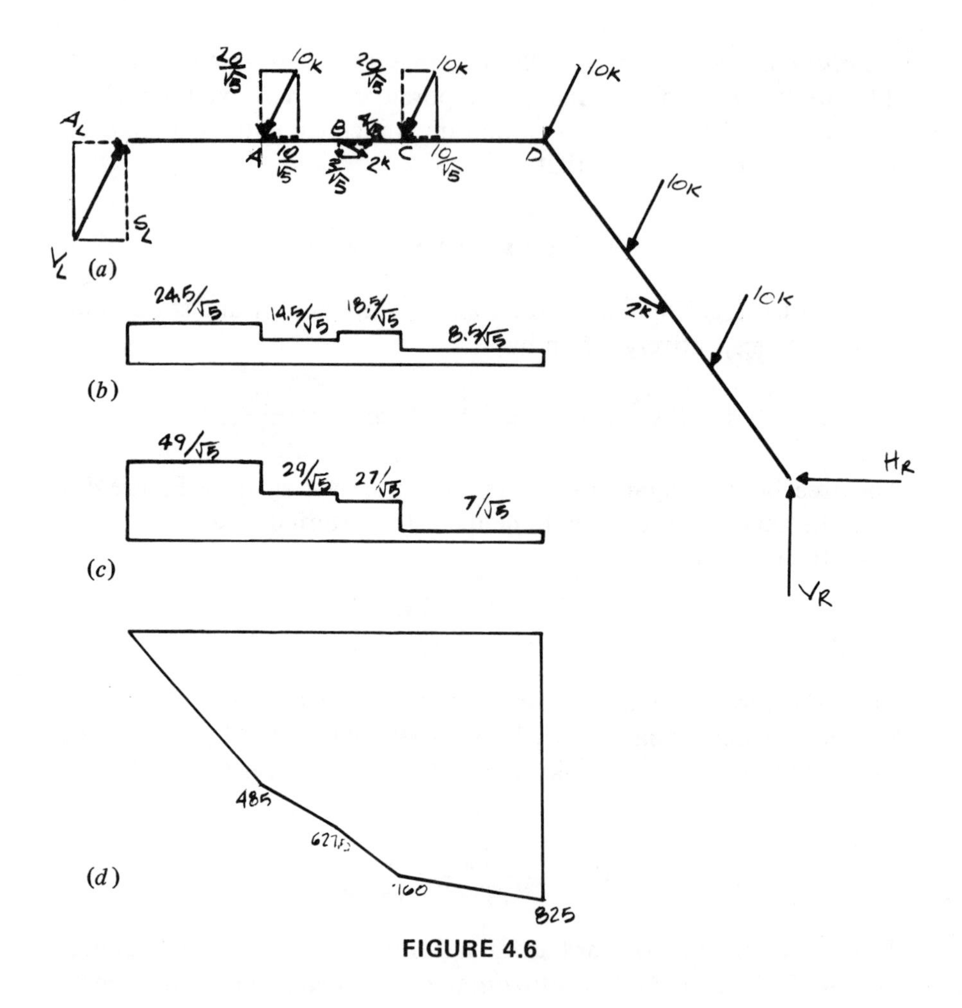

FIGURE 4.6

4.6. Following the solution to Problem 4.5, draw the axial force diagram for the right part of the structure (Figure 3.6*b*).

4.7. Draw the shear diagram for the left part of the structure (Figures 3.4*b* and 4.6*c*).

Solution. At the roller, the shear force is

$$S_L = \frac{24.5(2)}{(5)^{1/2}} = \frac{49}{(5)^{1/2}} \text{ k.}$$

This shear remains constant as far as point A, where the shear component $20/(5)^{1/2}$ k of the 10-k load reduces it to

$$S_A = S_L - \frac{20}{(5)^{1/2}} = \frac{29}{(5)^{1/2}} \text{ k.}$$

From point A to B, the shear remains constant. At point B, the shear component $2/(5)^{1/2}$ k of the horizontal load reduces the shear to

$$S_B = S_A - \frac{2}{(5)^{1/2}} = \frac{27}{(5)^{1/2}} \text{ k.}$$

This shear remains constant as far as point C, and it is there reduced to

$$S_C = S_B - \frac{20}{(5)^{1/2}} = \frac{7}{(5)^{1/2}} \text{ k,}$$

which remains constant from C to D (Figure 4.6*c*).

4.8. Draw the shear diagram for the right part of the structure (Figures 3.6*b* and 4.6*c*).

4.9. Draw the moment diagram for the left part of the structure (Figures 3.6*b* and 4.6*d*).

Solution. Only the shear components of external forces produce bending moments, since axial components lack lever arms around any section of the structure. Starting from the roller where the moment is zero, S_L produces clockwise moments that increase linearly up to point A. The axial distance from the roller to point A is

$$\frac{20}{\cos\alpha} = \frac{20(5)^{1/2}}{2} = 10(5)^{1/2} \text{ ft.}$$

The moment produced by S_L at A is then

$$M_A = -10S_L(5)^{1/2} = -485 \text{ k-ft.}$$

From point A on, the $20/(5)^{1/2}$-k shear component of the load produces counterclockwise moments detracting from the moments produced by S_L. At an axial distance x from the roller $[10(5)^{1/2} < x < 15(5)^{1/2}]$, the moment is

$$M(x) = -xS_L + [x - 10(5)^{1/2}]\frac{20}{(5)^{1/2}},$$

which shows a linear variation of moments from A to B (Figure 4.6d). Accordingly,

$$M_B = -15(5)^{1/2} S_L + [5(5)^{1/2}] \frac{20}{(5)^{1/2}} = -727.5 + 100 = -627.5 \text{ k-ft.}$$

Following similar reasoning, we realize that the moment also varies linearly from B to C and from C to D. At points C and D, values of the moment are

$$M_C = -20(5)^{1/2} S_L + [10(5)^{1/2}] \frac{20}{(5)^{1/2}} + [5(5)^{1/2}] \frac{2}{(5)^{1/2}}$$

$$= -970 + 200 + 10 = -760 \text{ k-ft.}$$

$$M_D = -30(5)^{1/2} S_L + [20(5)^{1/2}] \frac{20}{(5)^{1/2}} + [15(5)^{1/2}] \frac{2}{(5)^{1/2}}$$

$$+ [10(5)^{1/2}] \frac{20}{(5)^{1/2}} = -1455 + 400 + 30 + 200 = -825 \text{ k-ft.}$$

4.10. Draw the moment diagram for the right part of the structure (Figures 3.6b and 4.6d).

4.11. Starting from the left end, find axial force A, shear force S, and bending moment M at the midspan of the structure (Figure 3.7).

4.12. Starting from the right end, find axial force A, shear force S, and bending moment M at the midspan of the structure (Figure 3.7).

4.13. From the left, calculate axial force A, shear force S, and bending moment M at the midspan of the structure (Figure 3.8b).

Solution. The axial force has the direction of the local tangent to the structural axis, which is horizontal at midspan. Thus, summing all horizontal external forces from the left end to midspan, we obtain

$$A = R_L \cos 45° = 20(0.707) = 14.14 \text{ k.}$$

The shear force has the direction of the local radius of the structural axis, which is vertical at midspan. Thus, summing all vertical external forces from the left end, we obtain at midspan

$$S = R_L \sin 45° - 10 = 14.14 - 10 = 4.14 \text{ k}.$$

At midspan, the shear force diagram has a 28.4-k step due to the local loading force. The midspan bending moment is

$$M = M_L + 20R_L \cos 45° - 40R_L \sin 45° + 10(20)$$

$$= -710 + 282.8 - 565.6 + 200 = -792.8 \text{ k-ft}.$$

4.14. From the right, calculate axial force A, shear force S, and bending moment M at the midspan of the structure (Figure 3.8*b*).

4.15. Starting from the left end, find the midspan internal forces and moment (Figure 3.9).

4.16. Starting from the right end, find the midspan internal forces and moment (Figure 3.9).

4.17. Draw a diagram of axial forces on the frame (Figure 3.10*b*).

Solution. From the vertical link to point C, the axial force vanishes everywhere for lack of horizontal external forces. From the slide to point D, the axial force coincides with R_R, which is the only external force with the direction of the structural axis. At point D, the 5-k axial component of the 10-k horizontal load decreases the value of the axial force to 15 k. This value remains constant from point D to C for lack of axial external forces (Figure 4.7*a*).

4.18. Draw a diagram of shear forces on the frame (Figure 3.10*b*).

Solution. At the section constrained by the vertical link, the shear coincides with $V_L = 53.32$ k. This value remains constant as far as point A, where the local load - 18 k reduces it to 35.32 k. From A to B, the shear remains constantly equal to 35.32. At point B, the - 18-k load changes the value of the shear to 17.32 k, which remains unchanged as far as point C. From the slide to point D, the shear vanishes everywhere for lack of external shearing forces. At point D, the 8.66 component of the local 10-k load coincides with the shear force, which remains constant from D to C (Figure 4.7*b*).

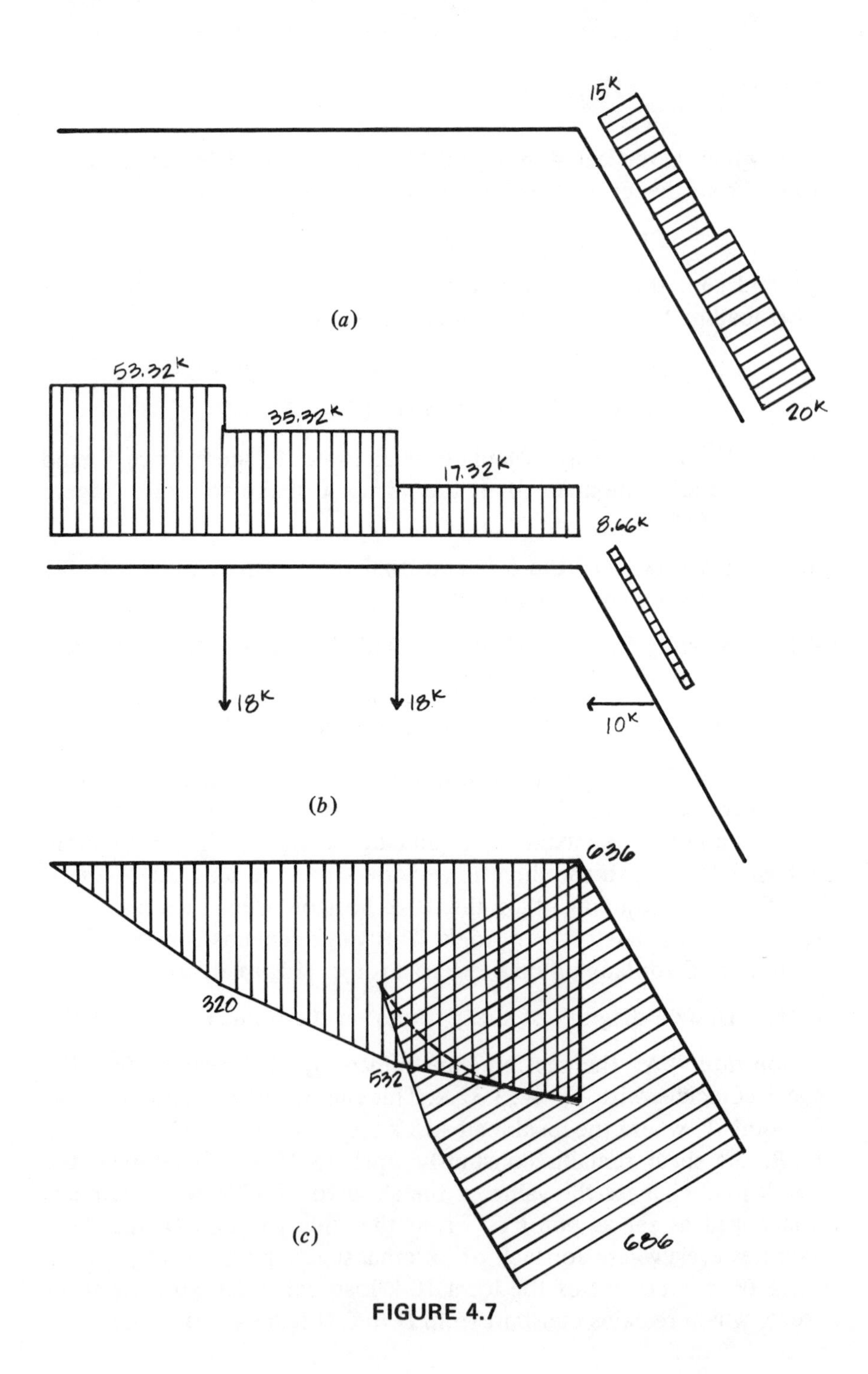

FIGURE 4.7

4.19. Draw the moment diagram on the frame (Figure 3.10*b*).

Solution. The moment on the beam increases linearly from zero at the section constrained by the link to $6V_L$ at point A. Thus,

$$M_A = -6(53.32) = -319.9 \text{ k-ft.}$$

At any point x between A and B ($6 < x < 12$ ft), the moment is

$$M(x) = -xV_L + (x - 6)18,$$

a linear function of the x corrdinate. Thus,

$$M_B = -12(53.32) + 6(18) = -531.8 \text{ k-ft.}$$

The moment diagram is also linear between B and C, and at point C, it attains the value

$$M_C = -18(53.32) + 12(18) + 6(18) = -635.8 \text{ k-ft.}$$

From the slide to point D, the moment remains equal to $M_R = +686$ k-ft (counterclockwise) for lack of moment inducing external forces. From D to C, the 10-k horizontal load produces clockwise moments to be subtracted from M_R (Figure 4.7*c*). At point C,

$$M_C = M_R - 10(5) = 686 - 50 = 636 \text{ k-ft.}$$

The complete moment diagram is shown in Figure 4.7*c*.

4.20. Draw the axial force, shear force, and bending moment diagram on the structure; solve Problem 3.8 first (Figure 3.11).

4.21. Draw the diagrams of shear forces and bending moments on the beam; see Problem 3.9 (Figure 3.12*b*).

4.22. Draw the diagrams of shear forces and bending moments on the beam; solve Problem 3.10 first (Figure 3.13).

View of a typical roof beam in the project in Figure 5.11. Structures by Riccardo Morandi. Photograph courtesy of Professor Morandi, Rome.

FIVE

FLEXURAL MEMBERS: BEAMS

5.1. INTERNAL FORCES OF BEAMS: NUMERICAL EVALUATION

When the internal forces of a structural element consist of only the shear force and bending moment, the element is said to have beam behavior. This is usually the case with horizontal structural components subject to gravity loads. Depending on their position in the structural hierarchy, these components are called joists, beams, or girders. The reactions of the constraints of statically determinate beams are evaluated according to the procedure outlined in Chapter 3; that is, by setting up and solving equations of equilibrium of the free body. Shear forces and bending moments at any cross section S of the beam are evaluated according to the procedure outlined in Chapter 4; that is, by adding algebraically all contributing forces or moments from one end of the beam to section S.

5.1.1. Beams without Shear or Moment Releases

The girder in Figure 5.1 supports three beams that discharge concentrated loads on it as shown. The equilibrium of the free body is stated in the following equations:

$$\Sigma F_x = 0 = H_L ,$$

$$\Sigma M_A = 0 = -4(8) - 10(12) - 16(4) + 20V_R ,$$

$$\Sigma F_Y = 0 = V_L - 8 - 12 - 4 + V_R .$$

Solutions to the equations are:

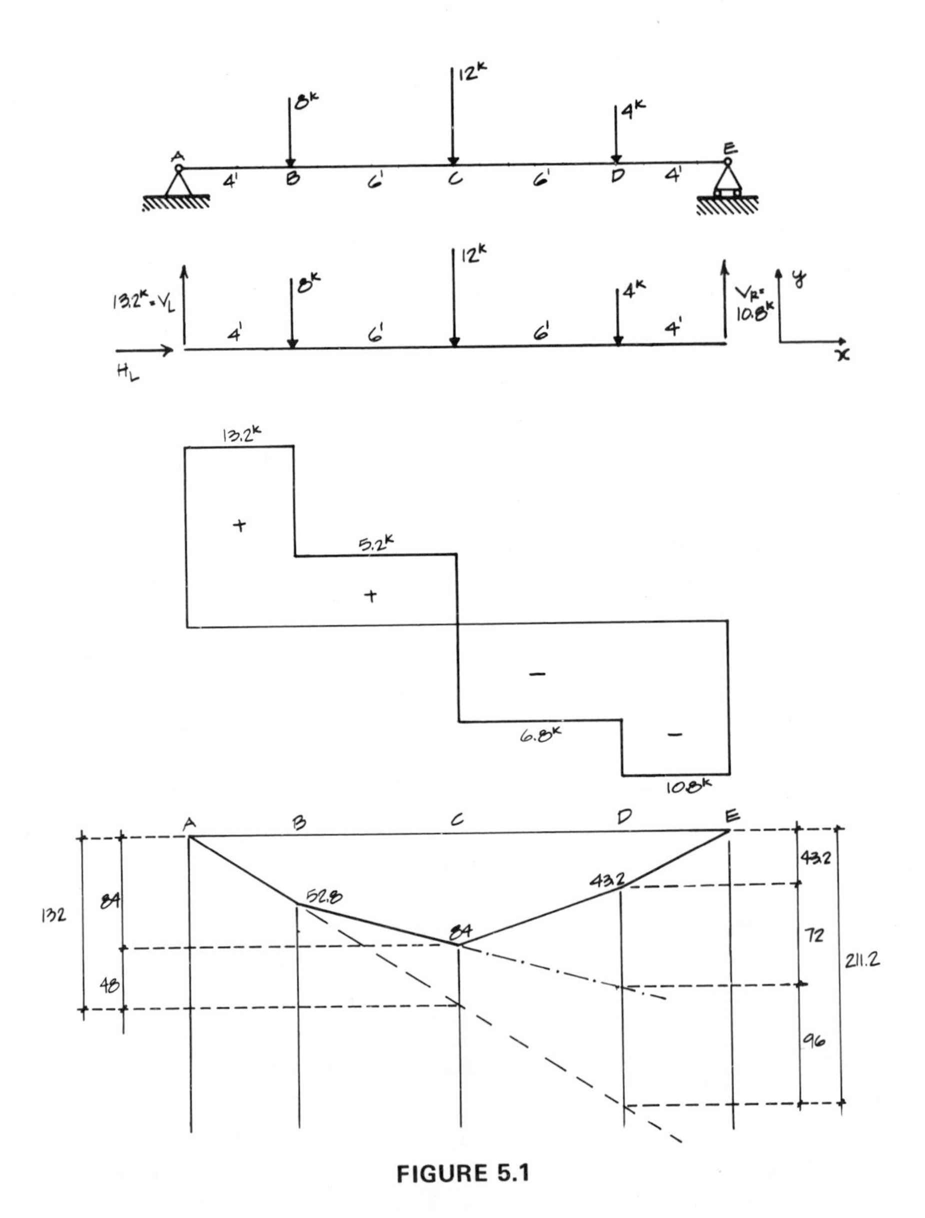
12k
8k
4k
A
E
4'
B
6'
C
6'
D
4'
12k
8k
4k
13.2k = VL
VR = 10.8k
y
x
HL
13.2k
+
5.2k
+
−
6.8k
−
10.8k
A
B
C
D
E
132
84
48
52.8
84
43.2
43.2
72
96
211.2

FIGURE 5.1

$$H_L = 0,$$

$$V_R = \frac{32 + 120 + 64}{20} = 10.8 \text{ k},$$

$$V_L = 24 - V_R = 13.2 \text{ k}.$$

The shear diagram is drawn by reasoning as follows: The shear force at point A is the reaction $V_L = +13.2$ k, to which nothing is added or substracted along segment AB of the beam axis. At point B, the -8-k concentrated load reduces the value of the shear force from $+13.2$ to $+5.2$ k. This value remains constant along segment BC of the beam axis. At point C, the -12-k external force changes the value of the shear force from $+5.2$ to -6.8 k, which remains constant along segment CD of the beam axis. At point D, the -4-k load increases the negative value of the shear from -6.8 to -10.8 k. At point E, the $+10.8$ reactive force neutralizes the -10.8-k shear force, and the shear force diagram returns to its base line (Figure 5.1).

The moment diagram is drawn according to the following considerations: Applied and reactive moments are zero at section A because the leftmost external force V_L lacks a level arm around A, and the hinge connection excludes a reactive moment (Figure 5.1). The internal moment M_A that balances the external moments is, therefore, zero at point A.

Proceeding to sections that are 1, 2, or 3 ft away from section A, the lever arm of the external force V_L, and therefore its moment, gradually increase. The increase is linear, since the moment doubles, triples, and quadruples with the lever arm. At point B,

$$M_B = 4V_L = 4(13.2) = 52.8 \text{ k-ft}.$$

External forces V_L and 8 k both contribute to determining the values of moments on beam sections along segment BC. Thus, the counterclockwise moments produced by V_L, shown in Figure 5.1 by the dotted diagram, are reduced by clockwise moments produced by the 8-k load. The solid moment diagram is the difference between the outer triangle with origin at A and the inner triangle with origin at B.

At point C,

$$M_C = 10(13.2) - 6(8) = 84 \text{ k-ft}.$$

Three external forces produce bending moments on beam sections along segment *CD*; they are reaction V_L and the 8 and 12-k concentrated loads. Figure 5.1 shows that the moments along the segment of the beam axis are obtained by subtracting the triangle with origin at *B*, produced by 8 k, and the triangle with origin at *C*, produced by 12 k, from the diagram with origin at *A*, produced by V_L. Thus,

$$M_D = 16(13.2) - 12(8) - 6(12) = 43.2 \text{ k-ft}.$$

Finally, at point *E*,

$$M_E = 13.2(20) - 8(16) - 12(10) - 4(3) = 0.$$

The value of M_E is in agreement with the requirement that the moment equal zero at a pinned section.

It is useful to practice calculating the moments M_D, M_C, M_B, and M_A, proceeding from the right to the left end. Examples worked out at the end of this chapter and the various problems proposed provide additional practice.

The moment diagram in Figure 5.1 is below the base line in accordance with the rule that suggests drawing the diagram on the side of the tension fibers. If tension due to bending in the beam occurs in the upper fibers (as in the case of cantilever beams), the moments are shown above the base line and vice versa. If this suggestion is followed, the shape of the moment curve resembles the shape of the beam axis after bending. For example, the curve of the beam axis in Figure 5.2*a* resembles the moment diagram in Figure 5.2*b*.

Shear forces and bending moments are naturally related to loads. They are also interdependent, since the moments are produced by the same external forces that are added to obtain the shear. Awareness of load-shear-moment relations is essential for a general understanding of structural designs and behavior. In order to investigate this interdependence, we resort to the free body diagram of an infinitesimal segment of a beam (Figure 5.3). For this slice of beam, we can write an equation of vertical force equilibrium, which is a mathematical statement of the physical evidence that the element does not move vertically. Similarly, we can state that the element is not rotating with an equation of moment equilibrium.

Neighboring sections of the beam exert a shear force V and a moment M on the element at the x-coordinate as well as a shear force $V + dV$ and a moment $M + dM$ at the coordinate $x + dx$. Incre-

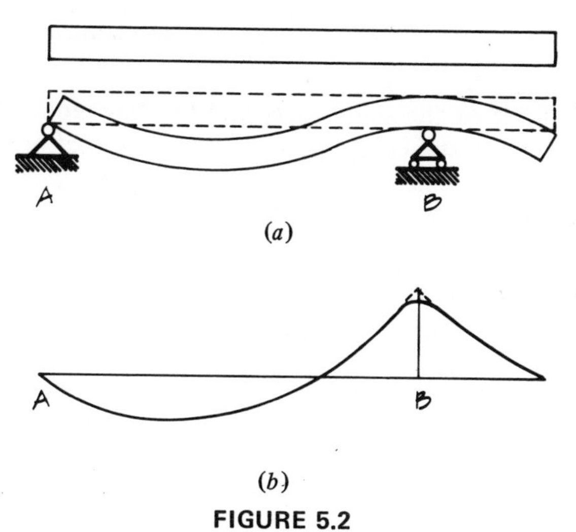

FIGURE 5.2

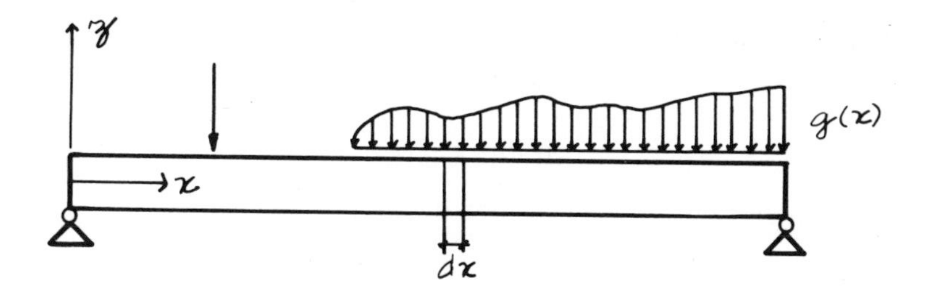

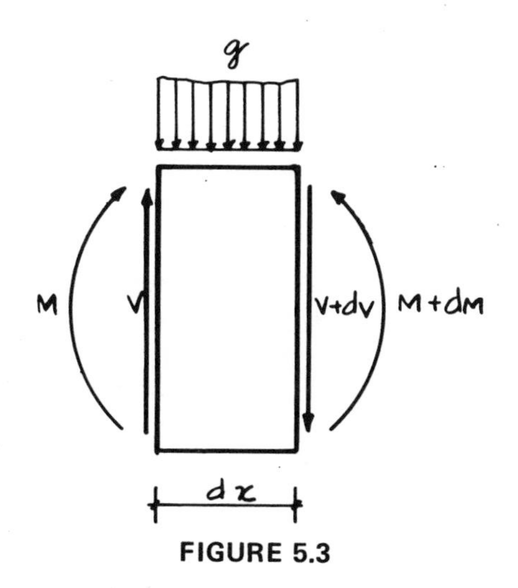

FIGURE 5.3

ments dV and dM of the shear force and bending moment are infinitesimal, since the distance dx between terminal sections of the element is infinitesimal. The free body diagram also shows the element's share of the load on the beam. The intensity g of the load can be considered constant along the infinitesimal distance dx. Then, the equation for vertical equilibrium is

$$V - g\,dx - (V + dV) = 0.$$

The equation of moment equilibrium around the center of the section at the coordinate $x + dx$ is

$$-M - V\,dx + g\,dx\,\frac{dx}{2} + (M + dM) = 0.$$

Eliminating equal and opposite terms from the equations and discarding the second-order infinitesimal $g(dx)^2/2$ from the equation of moments, we obtain

$$dV = -g\,dx,$$

$$dM = V\,dx.$$

After dividing by dx, the equations become

$$-g = \frac{dV}{dx},$$

$$V = \frac{dM}{dx}.$$

It is evident from these relations that shear force and bending moment always coexist, since shear is the rate of change of the moment along the beam axis. The case of constant bending moment and zero shear (the moment derivative) is unusual except at individual sections where the moment has extreme values.

Recalling that the ratio $\Delta M/\Delta x$ of the increment ΔM of bending moment and the distance Δx between two sections is the slope of the chord of the moment diagram (Figure 5.4), we can view the shear

$$V = \frac{dM}{dx} = \lim_{\Delta x \to 0} \frac{\Delta M}{\Delta x}$$

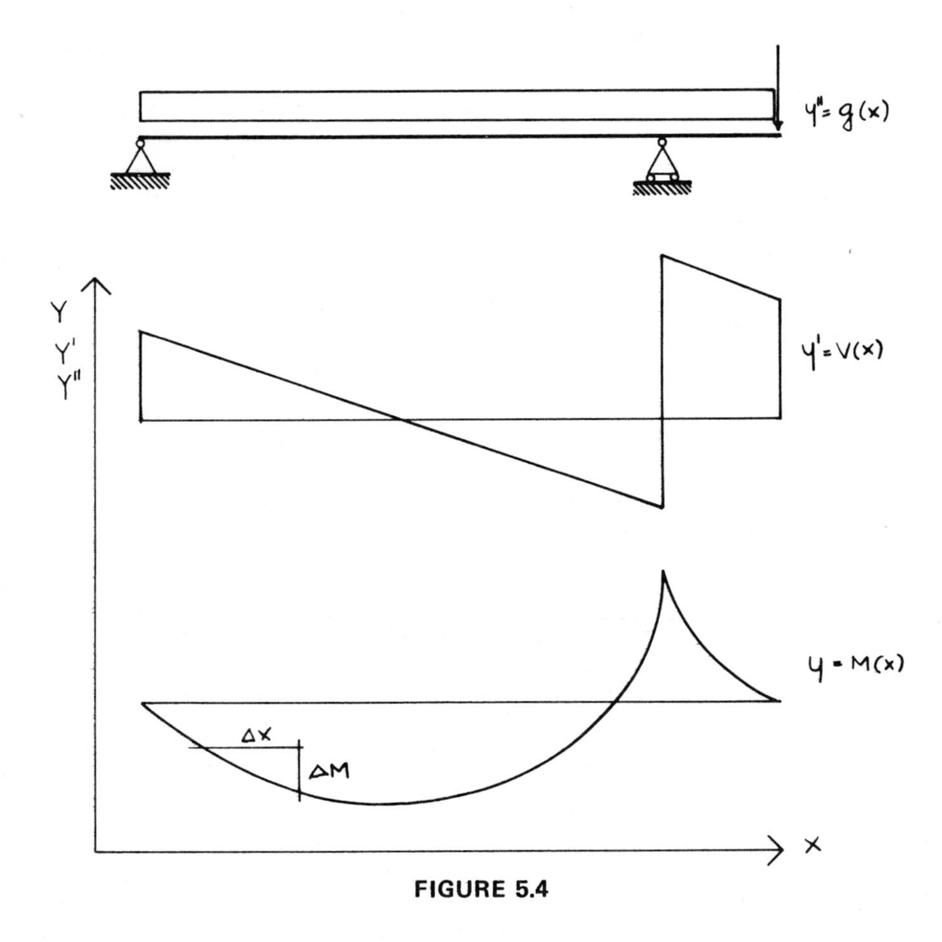

FIGURE 5.4

as the slope of the tangent of the moment diagram. At the sections where the moment curve attains a maximum value and then decreases, its slope, the shear, changes from positive to negative or vice versa.

It is possible, therefore, to single out sections where $V = 0$ and calculate directly the maxima of the moment at those sections. Considering, for example, the beam in Figure 5.5, we can find coordinate $\bar{x}$ of the section where the bending moment is maximum by setting the shear $V(\bar{x})$ equal to zero and solving for $\bar{x}$.

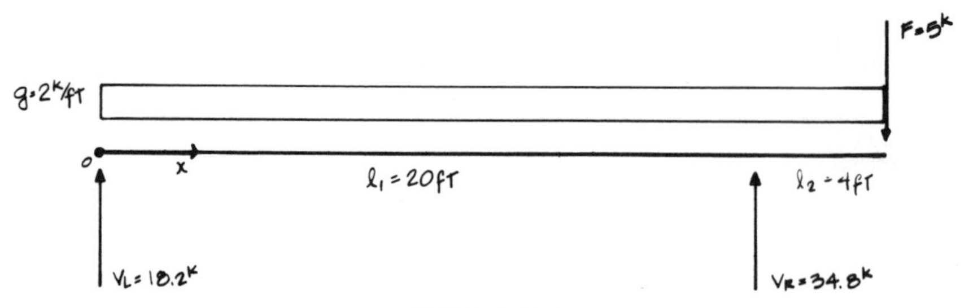

FIGURE 5.5

$$V(\bar{x}) = V_L - g\bar{x} = 18.2 - 2\bar{x} = 0$$

$$\bar{x} = \frac{V_L}{g} = \frac{18.2}{2} = 9.1 \text{ ft.}$$

Then,

$$M(\bar{x}) = V_L\bar{x} - g\bar{x}\frac{\bar{x}}{2} = V_L \frac{V_L}{g} - \frac{g}{2}\frac{V_L^2}{g^2}$$

$$M(\bar{x}) = \frac{V_L^2}{2g} = \frac{(18.2)^2}{4} = 82.8 \text{ k-ft.}$$

If the origin of the beam axis is taken at the right end of the span (Figure 5.6),

$$V(\bar{x}) = V_0 - g\bar{x} = (V_R - F - gl_2) - g\bar{x} = 21.8 - 2\bar{x} = 0$$

$$\bar{x} = \frac{V_0}{g} = \frac{21.8}{2} = 10.9 \text{ ft,}$$

then

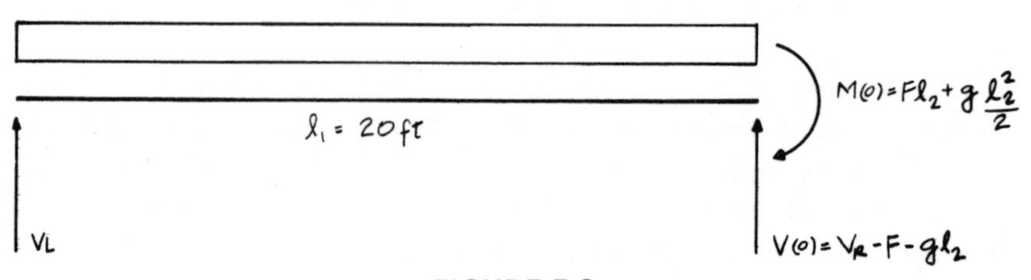

FIGURE 5.6

$$M(\bar{x}) = V_0\bar{x} - g\bar{x}\frac{\bar{x}}{2} - M_0 = V_0\frac{V_0}{g} - \frac{g}{2}\frac{V_0^2}{g^2} - M_0 = \frac{V_0^2}{2g} - M_0.$$

Since

$$M_0 = Fl_2 + gl_2\frac{l_2}{2} = 20 + 16 = 36 \text{ k-ft},$$

then

$$M(\bar{x}) = \frac{21.8^2}{4} - 36 = 82.8 \text{ k-ft}.$$

By using load-shear-moment relations, diagrams of internal forces can be sketched even before numerical values are calculated. For example, the shear diagram in Figure 5.7 is drawn by reasoning as follows: At point A, the shear V_A coincides with reaction V_L. From point A to point B, the load $g = -dV/dx$ increases linearly.

The derivative g of the shear is the slope of the shear curve, a function of x 1° lower than the shear itself. That is, if the shear curve along segment AB of the beam is the parabola $V(x) = ax^2 + bx + c$, then the load g along AB is the straight line

$$g(x) = \frac{dV}{dx} = 2ax + b.$$

Therefore, we draw a parabolic shear curve from V_A to V_B, which has zero slope at A ($g_a = 0$) and an increasing slope that reaches the value $g_{1\,\max}$ at B.

The load decreases linearly from $g_{1\,\max}$ to zero along the beam segment BC. Accordingly, we draw a parabola with slopes decreasing from $g_{1\,\max}$ at B to zero at C. The point of the shear curve where the slope attains its greatest value $g_{1\,\max}$ and then starts decreasing is a point of inflexion. The segment CD of the beam axis is not loaded, therefore, the shear V_c remains constant as far as point D. It can also be said that along CD, where a load curve does not exist, the shear curve must have the lowest degree, such as $y' = V_c x^0 = V_c$, a constant.

At point D, external force F is subtracted from V_c to obtain V_D. The shear remains constant from D to E. At point E, external force V_R is added to V_D to obtain V_E. From E to F, the load diagram is the curve $y'' = g_2 x^0$, a zero-degree curve, a constant. The shear dia-

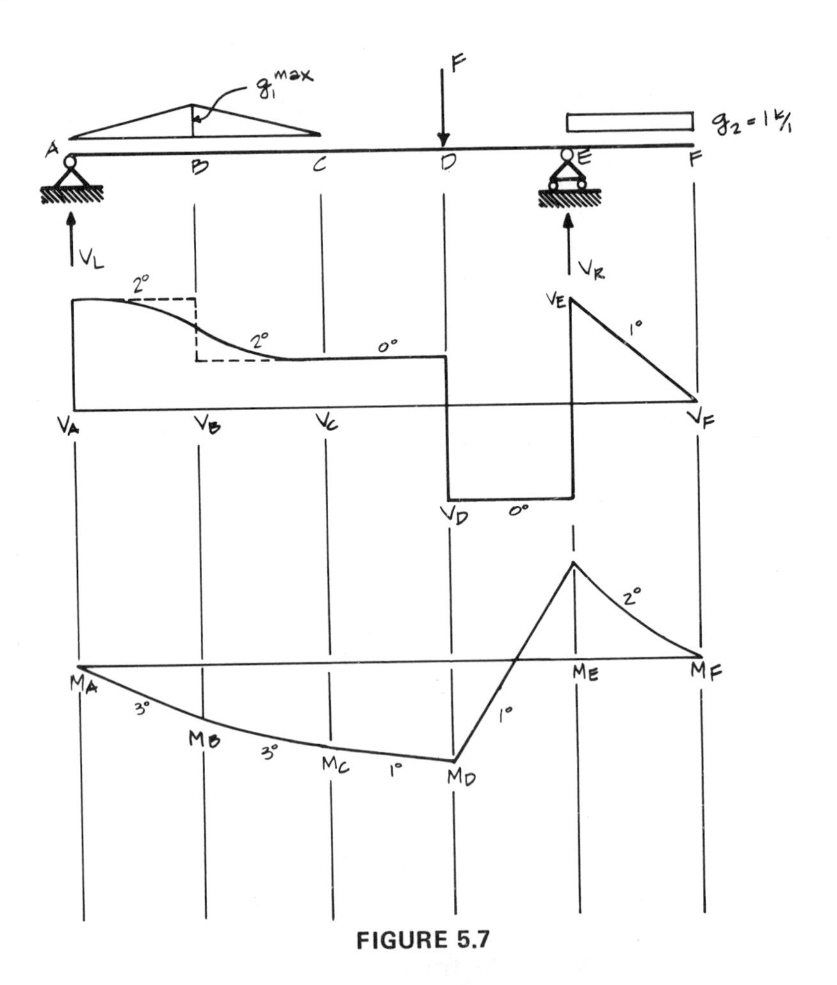

FIGURE 5.7

gram is a curve one degree higher, a straight line with slope g_2. At point F, the shear is zero, since this end of the beam is not subject to any external force.

To draw the moment diagram, we reason as follows: The pin end at point A requires $M_A = 0$. The shear $V = dM/dx$ is a decreasing, second-degree curve from A to C. The moment along this segment of beam axis is therefore a curve one degree higher (third degree), with slopes decreasing from V_A to V_C.

From C to D, the shear is the curve of zero degree, $y' = V_C x^0 = V_C$, a constant. The moment is, then, a first-degree curve (straight

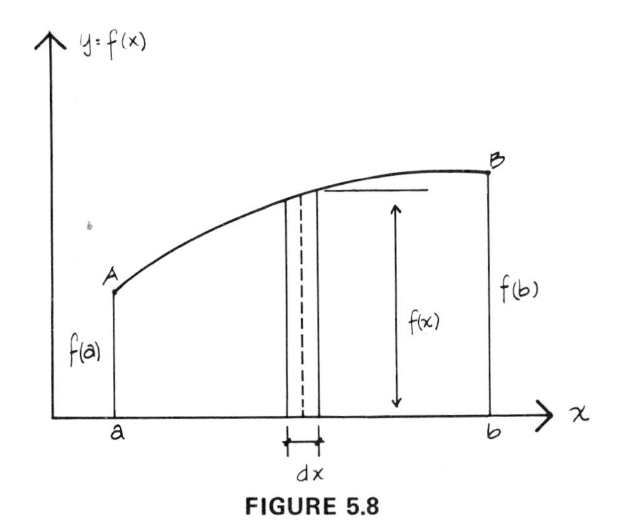

FIGURE 5.8

line with slope V_C). At point D, there is a sign reversal and a change in the value of the shear. Thus, the moment slope changes in value and sign. From E to F, the shear decreases linearly (first degree) from V_E to $V_F = 0$, and the moment is, accordingly, a parabola (second degree), with slopes decreasing to zero at the beam end F.

The load-shear-moment relations, written in integral rather than differential form, yield values for the internal forces in geometric terms. Recalling that the integral $\int_a^b f(x)\,dx$ is the sum of the infinite number of infinitesimal rectangles $f(x)\,dx$ under curve AB (Figure 5.8), the shear at a point x of the beam axis

$$V(x) = C_1 + \int g(x)\,dx$$

can be viewed as the algebraic sum of the area $\int g(x)\,dx$ under the load curve and of the concentrated external forces C_1 from the origin to x. Similarly, the moment at a point x of the beam axis

$$M(x) = C_2 + \int V(x)\,dx$$

can be viewed as the algebraic sum of the area $\int V(x)\,dx$ under the shear curve and of the concentrated moments C_2.

Considering, for example (Figure 5.9), span AE of the beam in

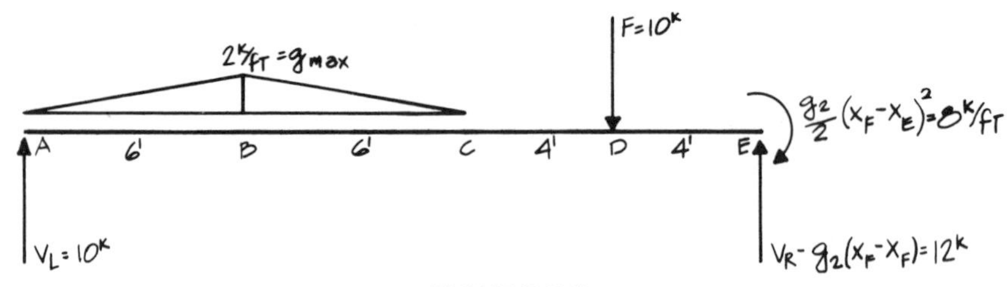

FIGURE 5.9

Figure 5.7, we can calculate the internal forces at intermediate points B, C, and D of the beam axis as follows:

$$V_B = C_1 - \int g(x)\,dx = V_L - g_{1\,\mathrm{max}} \frac{(x_B - x_A)}{2},$$

$$V_B = 10 - 2(3) = 4\text{ k},$$

$$V_C = V_L - g_{1\,\mathrm{max}} \frac{(x_C - x_A)}{2} = V_B - g_{1\,\mathrm{max}} \frac{(x_C - x_B)}{2},$$

$$V_C = 10 - 2(6) = -2\text{ k},$$

$$V_D = V_L - g_{1\,\mathrm{max}} \frac{(x_C - x_A)}{2} - F,$$

$$V_D = 10 - 2(6) - 10 = -12\text{ k},$$

$$M_B = C_2 + \int V(x)\,dx = 0 - V_A(x_B - x_A) + \tfrac{1}{3}(V_A - V_B)(x_B - x_A),$$

$$M_B = -10(6) + \tfrac{6}{3}(6) = -48\text{ k-ft},$$

where the last term is the area above the parabolic segment,

$$M_C = M_B - V_C(x_C - x_B) - \tfrac{1}{3}(V_B - V_C)(x_C - x_B),$$

$$M_C = -48 - [-2(6)] - \tfrac{1}{3}[4 - (-2)]6 = -48\text{ k-ft},$$

$$M_D = M_C - V_C(x_D - x_C),$$

$$M_D = -48 - [-2(4)] = -40\text{ k-ft},$$

$$M_E = M_D - V_D(x_E - x_D),$$

$$M_E = -40 - [-12(4)] = +8\text{ k-ft}.$$

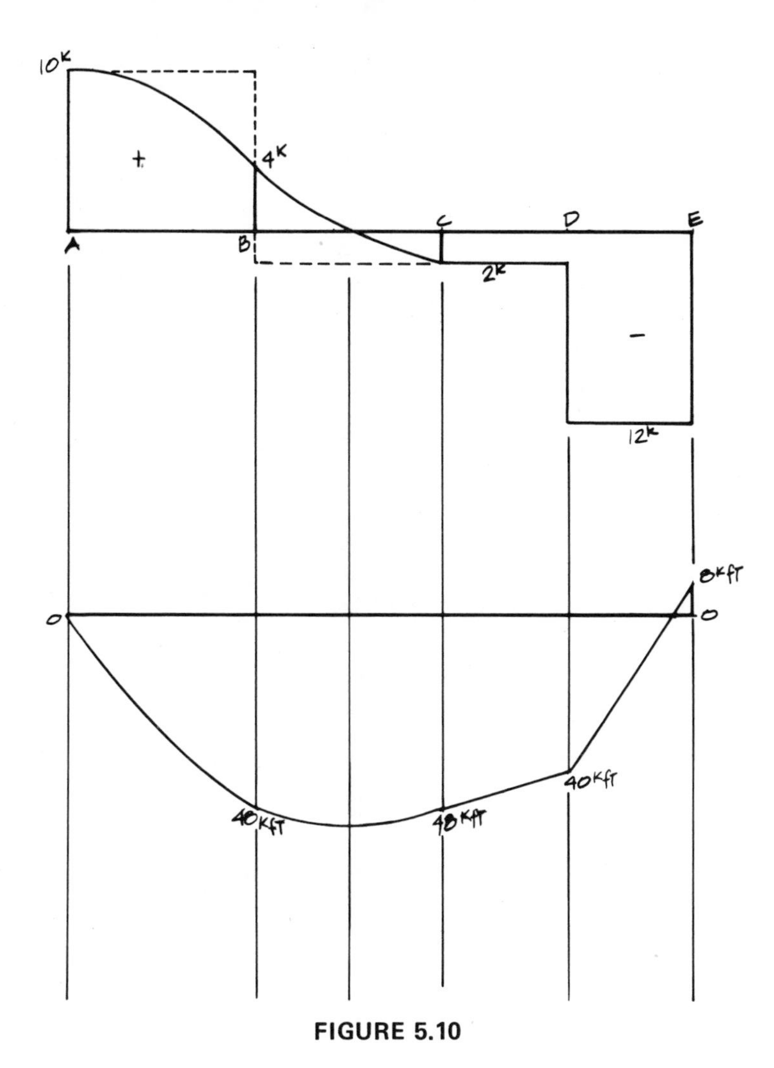

FIGURE 5.10

With these values and the shapes in Figure 5.7, the shear and moment diagrams (Figure 5.10) are now completely defined. If the bending moments are calculated from the right end, we obtain, again,

$$M_D = C_2 + \int V(x)\,dx = 8 - 12(4) = -40 \text{ k-ft},$$

$$M_C = M_D + V_C(x_D - x_C) = -40 - 2(4) = -48 \text{ k-ft},$$

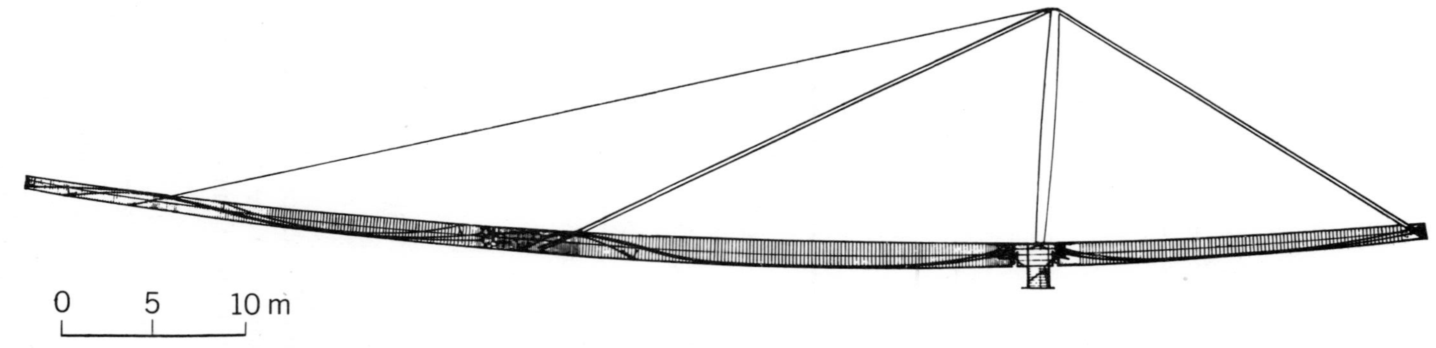

FIGURE 5.11. Structural drawing of a typical roof beam of an Alitalia hangar at Leonardo da Vinci Airport, Fiumicino, Rome. Structures by Riccardo Morandi, Rome. Photograph courtesy of Professor Morandi.

FIGURE 5.12. An elevation of the building in the chapter-opening photograph and Figure 5.11 after completion. Structures by Riccardo Morandi, Rome. Photograph courtesy of Professor Morandi.

$$M_B = M_C + V_C(x_C - x_B) - \tfrac{1}{3}(V_B - V_C)(x_C - x_B),$$

$$M_B = -48 - 2(6) + \tfrac{1}{3}[4 - (-2)]6 = -48 \text{ k-ft}.$$

5.1.2. Gerber Beams

When the distance between the terminal supports of a beam is too large to be covered by a single span, intermediate supports are provided, and the distance is divided into several spans short enough for using reasonably light beams. Standard structural elements of timber and laminated wood, rolled steel and precast concrete are usually available in limited lengths. Using intermediate supports eliminates the need for impractically deep and heavy members, but it leaves two problems unsolved: connecting elements of limited length to obtain a continuous beam and the redundancy of constraints resulting from a large number of supports.

Shear connections of steel framing members (Figure 5.13) are routinely made with simple erection procedures that resemble assembling a construction kit. A beam that is to be shear connected to a column is delivered from the factory with clip angles welded on both sides of the web near the mid-depth of each end section. The outstanding legs of the clip angles have bolt holes to match those

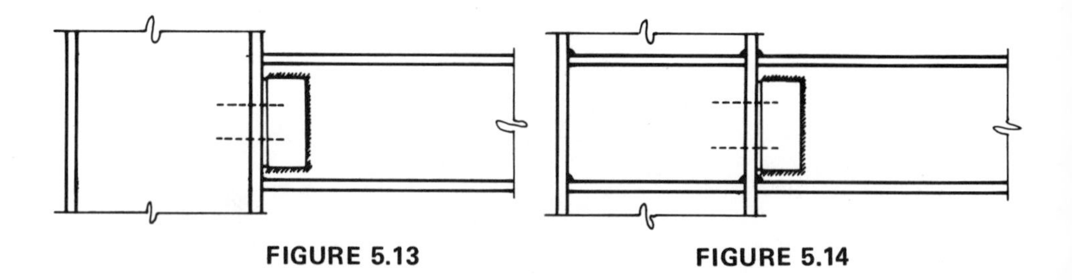

FIGURE 5.13

FIGURE 5.14

made in the factory on the flange or web of the column. During erection of the steel frame, one or two of the bolt holes are temporarily plugged with steel bars while bolts are machine tightened in the remaining holes. Finally, the steel bars are themselves replaced by bolts. In a beam-to-beam shear connection, the clip angles are replaced by flat steel plates.

Moment connections (Figure 5.14) must be made in the field. They are more complex and expensive because they require field welding, an operation that slows down the smooth flow of field erection. Field welding is done by operators working occasionally in less than comfortable positions in order to reach the base metal of the elements to be joined. Moreover, the connection that results can only be inspected ultrasonically rather than by X rays, as in the shop. For these reasons, structural designers prefer to avoid moment connections.

The redundancy of the structural scheme has the advantage of reducing bending moments produced by loads, but it presents, in addition to the inconvenience of field connections, stresses induced by distortions, as discussed in Chapter 3. The solution to both of these problems is using shear connections to join various parts of a beam, but with these connections placed at those points on the beam axis where bending moments under maximum load would vanish even if the beam were continuous.

As an example, a three-span, continuous beam is shown in Figure 5.15 with the diagram of bending moments induced by the loads. The beam has two redundant constraints and is therefore subject to unforeseeable internal forces and stresses in the event of foundation settlements. If, however, two internal hinges (moment releases or shear connections) are introduced at sections A and D, the beam becomes statically determinate. The moment diagram remains the

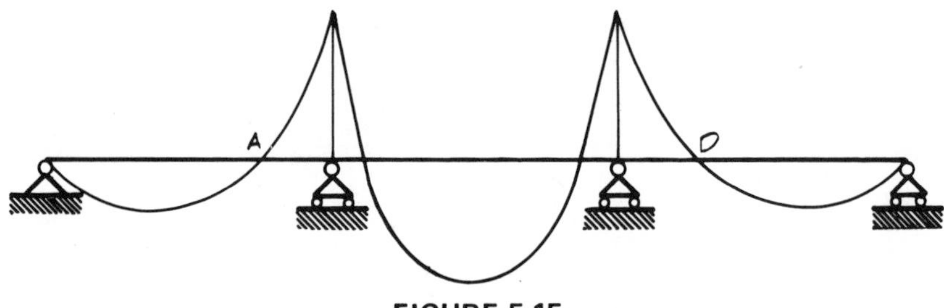

FIGURE 5.15

same, since M_A and M_D are zero in the continuous beam as well. The new scheme, therefore, retains the advantages of redundant structures while avoiding the disadvantages.

Statically determinate multi-span beams are often referred to as *Gerber beams*. The reactions of their external constraints and the interactions between solid pieces at the points of connection are obtained by the usual technique of drawing the free body diagram for each solid piece and writing the three equations of plane equilibrium. In these equations, reactions and interactions are the unknown forces or moments to be evaluated. For example, the beam in Figure 5.16 is made up of four monolithic parts joined by shear connections at points C, G, and K. The free body diagrams of the four parts are shown in Figure 5.16. In the absence of horizontal loads, the equations of equilibrium available are

$$\Sigma F_y = 0,$$

$$\Sigma M = 0.$$

If more than two unknown forces or moments appear in the free body diagram of one of the parts, two equations of equilibrium would not be sufficient to evaluate unknown reactions and interactions on that part of the beam. It is necessary, therefore, to start from the piece with the least number of unknown forces and moments, which, in this case, is AC. The equation of vertical equilibrium states

$$\Sigma F_y = 0 = -10 + V_C.$$

The equation of moment equilibrium around point C states

$$\Sigma M = 0 = -M_A + 10(4).$$

Therefore, $V_C = 10$ k, and $M_A = 40$ k-ft.

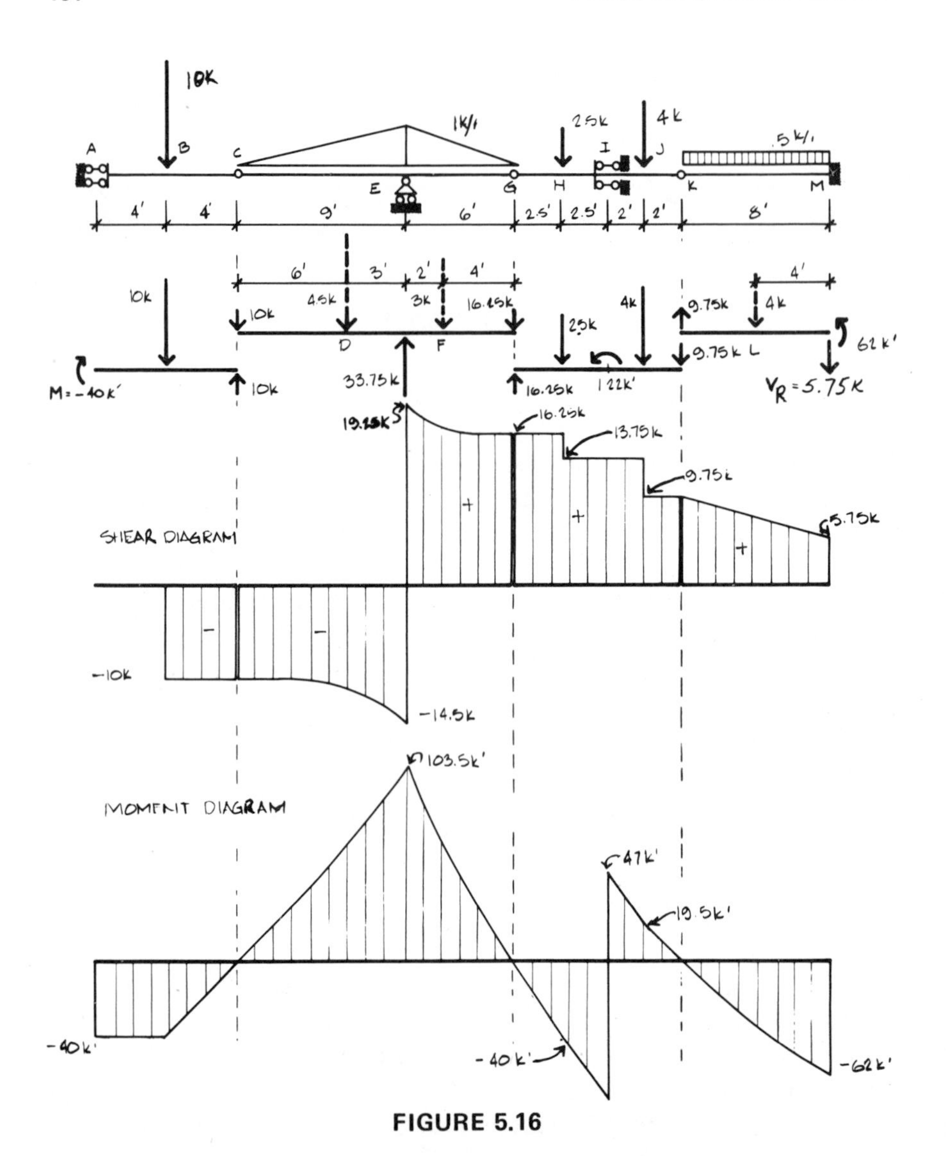

FIGURE 5.16

Similarly, for the piece CG, using G as the pivot point,

$$\Sigma M = 0 = +10(15) + 4.5(9) - 6V_E + 3(4),$$

from which

$$V_E = 33.75 \text{ k},$$

$$\Sigma F_Y = 0 = -10 - 4.5 + 33.75 - 3 - V_G.$$

Thus,

$$V_G = 16.95 \text{ k.}$$

For the part GK,

$$\Sigma F_Y = 0 = 16.25 - 2.5 - 4 + V_K,$$

from which

$$V_K = -9.75 \text{ k.}$$

The negative result indicates that the incorrect sign was assumed for V_K.

Using K as the pivot point,

$$\Sigma M = 0 = -9(16.25) + 2.5(6.5) + M_I + 4(2).$$

Thus,

$$M_I = 122 \text{ k-ft.}$$

After changing the sign originally assumed for V_K, the moment equilibrium of part KM is given by

$$\Sigma M = 0 = -4(4) - 8(5.75) + M_M.$$

Then,

$$M_M = 62 \text{ k-ft.}$$

With all external forces defined, the diagrams of shear force and bending moment are readily drawn as shown in Figure 5.16. Various examples and problems at the end of this chapter provide additional practice with Gerber beams.

5.2. REACTIONS AND INTERNAL FORCES OF BEAMS BY GRAPHICS

5.2.1. Single-Span Beams

The reactions of constraints, and shear and moment diagrams, can be obtained by graphic procedures as well as numerical ones from conditions of equilibrium. The brief review of graphic techniques in the appendix to this chapter is sufficient for understanding and applying these graphic methods.

Considering the beam in Figure 5.17, we first replace the distributed loads with their resultants, which equal in magnitude the areas of the various load distribution diagrams and are placed at their centroidal points. We then connect the lines of action of the loading forces with a string polygon (the pole P is arbitrary). The reactions V_L and V_R must provide force equilibrium and moment equilibrium for the system of loads. According to the graphic condition of force equilibrium, the sum of V_L and V_R must close the force polygon of the complete system of loads and reactions. Hence,

$$V_L + V_R = 4\text{–}1.$$

We must now ascertain how large a fraction of 4–1 is represented by V_L or V_R. The graphic condition of moment equilibrium provides the clue. The first and last sides of the string polygon, associated with a system of forces momentwise in equilibrium, must be colinear. Since reactions V_L and V_R form a system of forces momentwise in equilibrium with the loads, we can extend the string polygon of the loading forces to the full system of loads and reactions by adding a first side on the left of V_L and a last side on the right of V_R. These terminal sides of the extended string polygon must be colinear.

Next, we draw a line from pole P parallel to the sides added to the string polygon. This line divides $V_L + V_R$ at point O in the two separate reactions V_L = 0–1, preceding loading forces 1–2, 2–3, and 3–4 and V_R = 4–0, and following the loads. It is sufficient to read the drawings again by starting with 0–1 in the force polygon and with the side of the string polygon parallel to OP to realize that 0–1 and 4–0 satisfy both graphic conditions of equilibrium required for reactions V_L and V_R.

We now draw a horizontal base line through point O and project 0–1 horizontally from the force polygon onto the line of action of V_L and 1–2 onto its own line of action, and so on, with 2–3, 3–4, and 4.0 = V_R; thus, we obtain the shear diagram.

The moment diagram has already been drawn: It has the first and last sides of the string polygon as base line and the remaining sides of the string polygon are the diagram outline. To prove this, we recall that the moment of a system of forces around a given point Q is graphically represented by segment AB, intersected by the first and last sides of the string polygon on a line drawn through pivot point Q and parallel to the resultant of the system (see Appendix). Thus, we identify the moment at point Q of the beam axis, 12 ft from the left end, as follows: The external forces that contribute to moment M_Q are to the left of point Q; specifically, reaction V_L = 0–1 and loads 1–2 and 2–3. The string polygon associated with these three forces has four sides, of which OP is first and $3P$ is last. A line through section Q, parallel to the resultant 0–3 of external forces 0–1, 1–2, and 2–3, is cut at point A by OP and at point B by $3P$. AB represents the moment of 0–3 around section Q (the bending moment M_Q) in the scale

$$S_F S_D PH = (3\text{k/in.})(3\text{ ft/in.})(3\text{ in.}) = 27\text{ k ft/in.}$$

Reasoning similarly for any point Q on the beam axis, we realize that the string polygon is the moment diagram on base line OP.

The same conclusion can be reached on the basis of the integral formula for the moment $M(x)$ at the x-coordinate of the beam axis

$$M(x) = C_2 + \int C_1\, dx + \int dx \int g(x)\, dx.$$

The physical meaning of the three terms for $M(x)$ is: C_2 accounts for external moments, integral $\int C_1\, dx$ accounts for the moments of concentrated external forces, and the double integral $\int dx \int g(x)\, dx$ accounts for the moments of distributed loads.

The string polygon is the graphic integral of the concentrated loads, and the double integral of the distributed loads (their first integration is performed in replacing load distribution diagrams with their areas). The base line associated with the string polygon is the graphic equivalent of the numerical evaluation of the integration constants C_1 and C_2, according to moment boundary conditions. Hence, the string polygon and its base line are the graphic equivalent of $M(x)$.

The moment diagram in Fig. 5.17 can be drawn with a horizontal base line using a new pole P' on the horizontal line that divides V_L from V_R. If the distance $P'O$ equals the original polar distance PH, the moment scale remains the same. The new string polygon has its

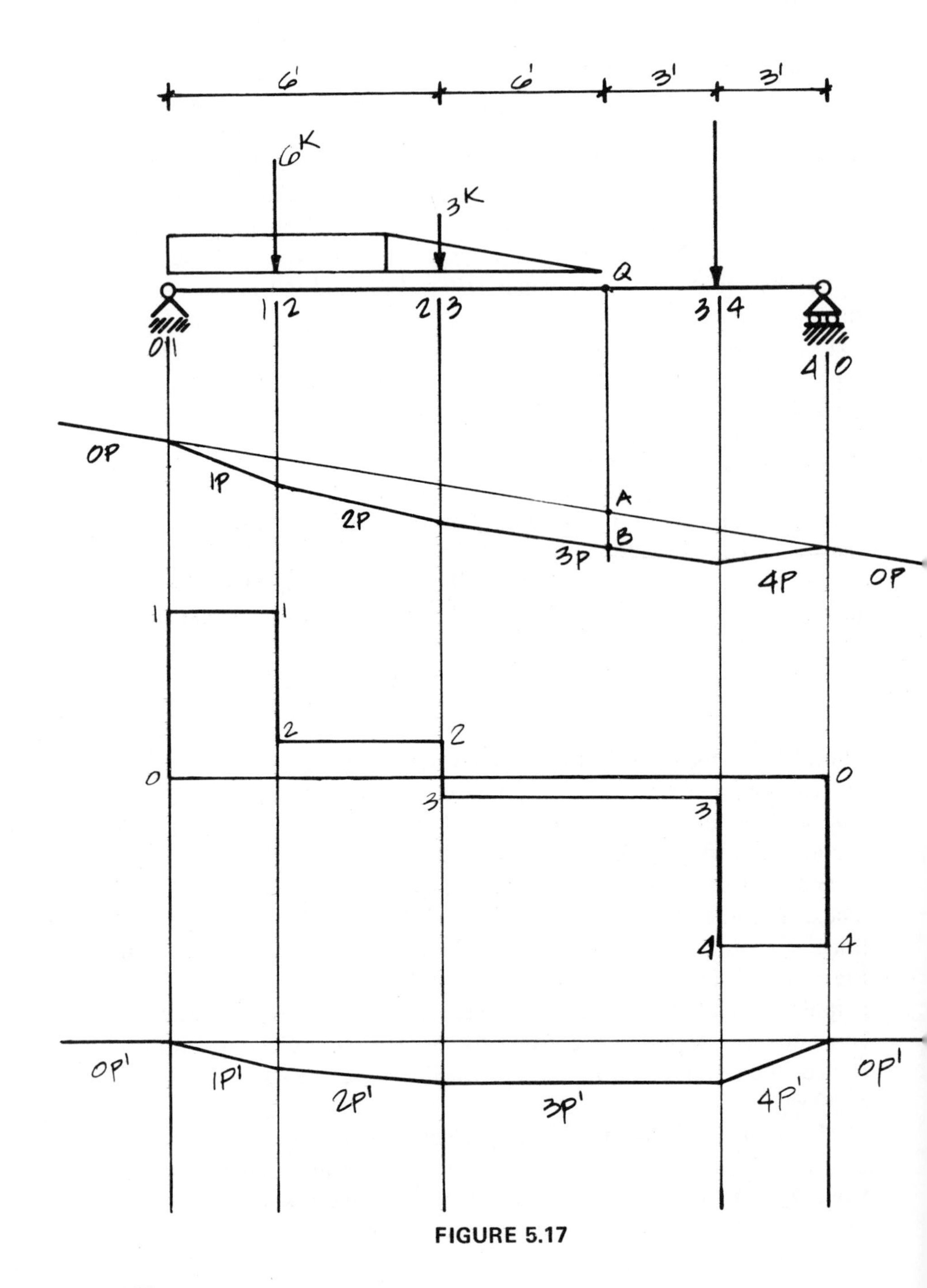

6'
6'
3'
3'
6K
3K
Q
0
1
1
2
2
3
3
4
4
0
OP
1P
2P
3P
4P
OP
A
B
1
1
2
2
0
0
3
3
4
4
OP'
1P'
2P'
3P'
4P'
OP'

FIGURE 5.17

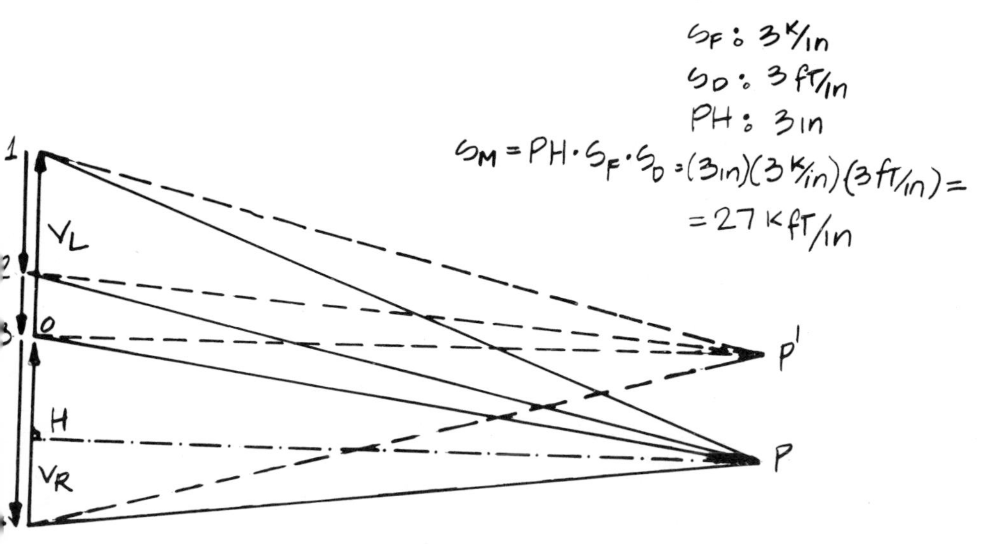

FIGURE 5.17. *(Continued)*

first-last side (base line) parallel to $P'O$, the second, third, fourth, and fifth sides parallel, respectively, to $1P'$, $2P'$, $3P'$, and $4P'$.

5.2.2. Multi-Span Beams

A span of a multi-span beam is a part of it extending between two external constraints. Overhangs extending from a constrained section to a free end are also considered separate spans. When a flexural member has several spans (Figure 5.18), each of them has different boundary conditions used in drawing the base line of the moment diagrams. For example, the first span AE of the beam in Figure 5.18 must have a base line that keeps the value of the moment constant from A to B, where the shear $V = dM/dx$ equals zero. The base line must also respect the condition that the moment equal zero at point C, where the hinge provides a moment release.

The base line of the second span EI must respect the condition that moment M_E at the left boundary be the same as that found at the right boundary of the span AE. The moment diagram must also change sign at the hinged section G. The base line of the third span

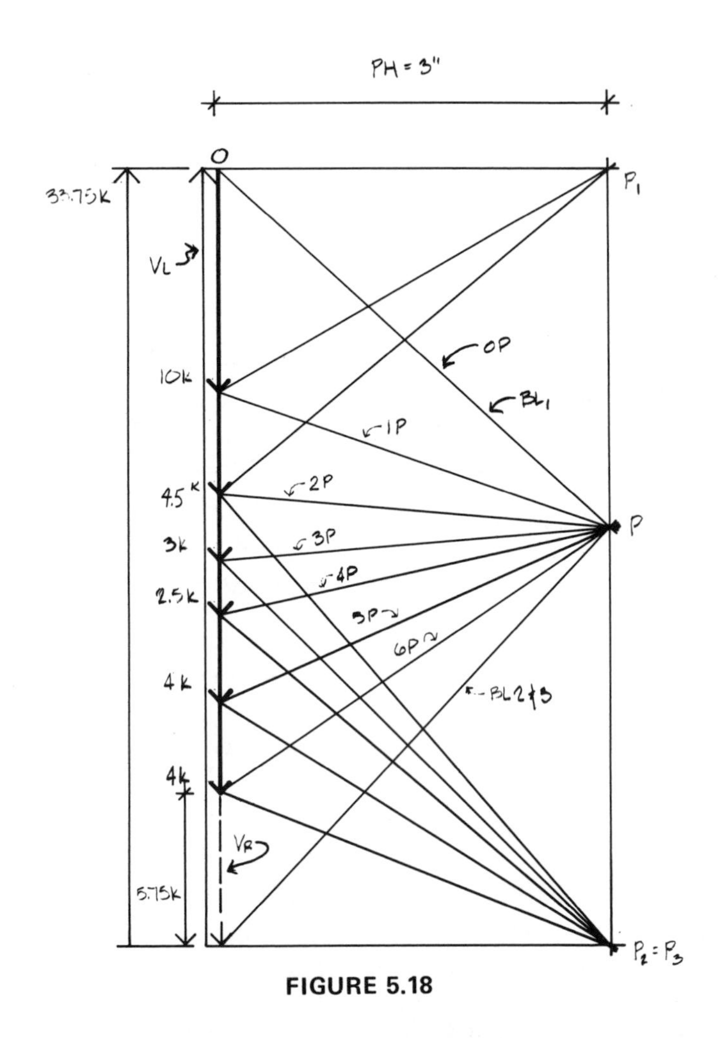

FIGURE 5.18

IM must cross the string polygon at section *K*, where the hinge requires zero moment, and it must be parallel to the base line of span *EI*. In fact, the external constraint that divides span *EI* from *IM* does not react with a vertical force. The shear on the left of section *I* is thus the same as the shear on the right of *I*. The equivalent geometric statement is that the moment curve on the left and right sides of point *I* (the fifth side $4P$ of the string polygon) has the same slope $V = dM/dx$ on the base line. It is clear, then, that each span must have its individual base line.

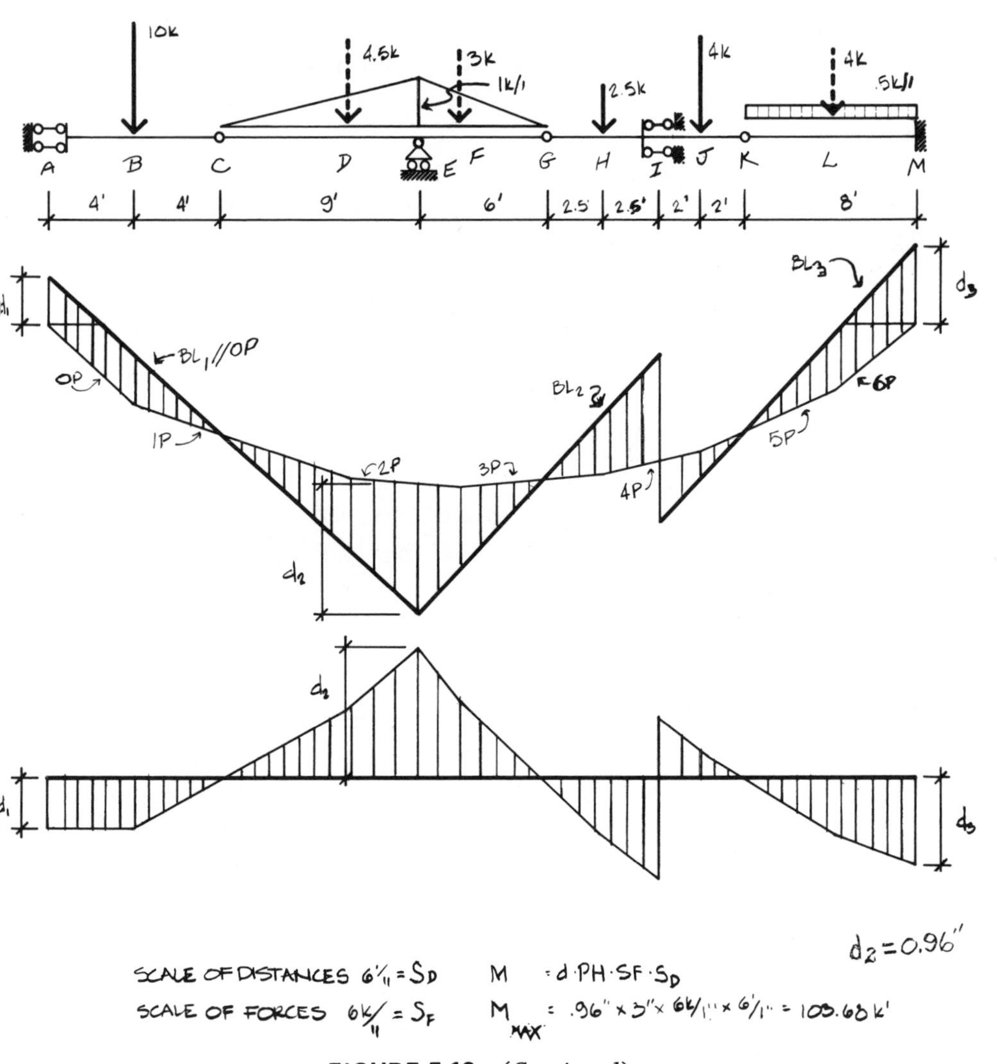

FIGURE 5.18. *(Continued)*

In Figure 5.18, the base lines of spans *AE*, *EI*, and *IM* have been drawn according to these specifications (boundary conditions) and labeled BL_1, BL_2, BL_3, respectively. The shear forces at the left and right ends of each span are found by the same approach used for reactions V_L and V_R of the single-span beam in Figure 5.17; that is, dividing the equilibrant of the span's load with a line from pole *P* parallel to the base line of the span. Thus, drawing the parallel *OP* to BL_1 from *P*, we divide the 14.5-k equilibrant of the loads on the first span into the right-end, 14.5-k shear force and the left-end, 0-k shear force.

Drawing the parallel to BL_2 from *P*, we divide the 5.5-k equilibrant of the loads on the second span into the downward, 13.75-k, right-end shear and the upward, 19.25-k left-end shear. Finally, drawing the parallel to BL_3 from *P*, we divide the 8-k equilibrant of the loads on the third span into the downward, 5.75-k, right-end shear and the upward, 13.75-k, left-end shear.

Shear forces on the left and right sides of an intermediate support are the shares for the left and right spans of the support's reaction. For example, the roller at point *E* provides a 14.5-k, upward shear force for the span *AE* on its left and an upward, 19.25-k shear force for the span *EI* on its right. The roller therefore provides a total upward reaction of 33.75 k. The slide at point *I* provides a downward, 13.75-k shear force for the span *EI* on its left and an upward, 13.75-k shear force for the span *IM* on its right. The slide therefore reacts with a total vertical force of *O* k or, rather, as we know, does not react vertically.

The shear diagram for each span is obtained by drawing the base line from the point that divides the equilibrant of the load of the span into left-end and right-end shear forces and then horizontally projecting end shear forces and loads from the force polygon. The bending moment diagram can be drawn with a horizontal base line common to all the spans by projecting external forces from the new poles P_1, P_2, P_3 (Figure 5.18), each of which belongs to one span, and it is taken on the horizontal line that divides the equilibrant of a span's loads into left- and right-end shear forces.

APPENDIX

In this appendix, we discuss the graphic evaluation of the resultant and the moment of a system of coplanar forces and graphic conditions for equilibrium. The resultant of a system of forces is a unique force equivalent to the system. To be equivalent, this force must produce the same translation and rotation as those produced by the

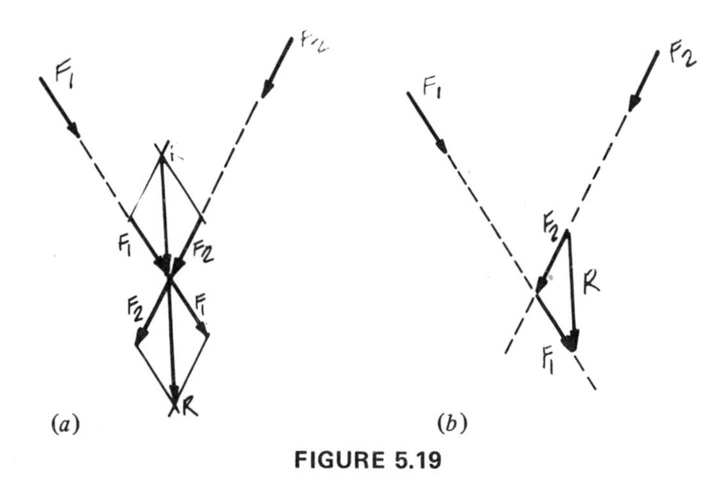

FIGURE 5.19

original system. Therefore, the resultant must be the vectorial sum of the forces of the system (components), and it must have the same moment around any point in space as that of the original forces.

The resultant of a system of two forces is obtained graphically by the *parallelogram rule* (Figure 5.19*a*). The two forces are moved along their lines of action so that they converge on the same point or diverge from it. Then a parallelogram is drawn by using the given forces as initial sides. The diagonal of the parallelogram is the resultant of the two forces. This procedure can be used to combine the resultant of the forces with a third force and the new resultant with a fourth force, and so on. If the system contains a large number of forces, evaluating the resultant by this approach is tedious and inefficient.

A similar rule, the *triangle rule*, gives the resultant of two forces by moving them along their line of action to a common point, as in Figure 5.19*b*. The third side of a triangle, whose initial sides are the given forces, is their resultant. The triangle rule does not give the correct line of action for the resultant, because that force must contain the common point to which both components are applied.

The resultant of a system with several coplanar forces can be found more synthetically by using the *string polygon* technique. The forces, shown in Figure 5.20*a*, are labeled 0–1, 1–2, and 2–3 and drawn again in a force polygon in Figure 5.20*b*. Points 0, 1, 2, . . . of

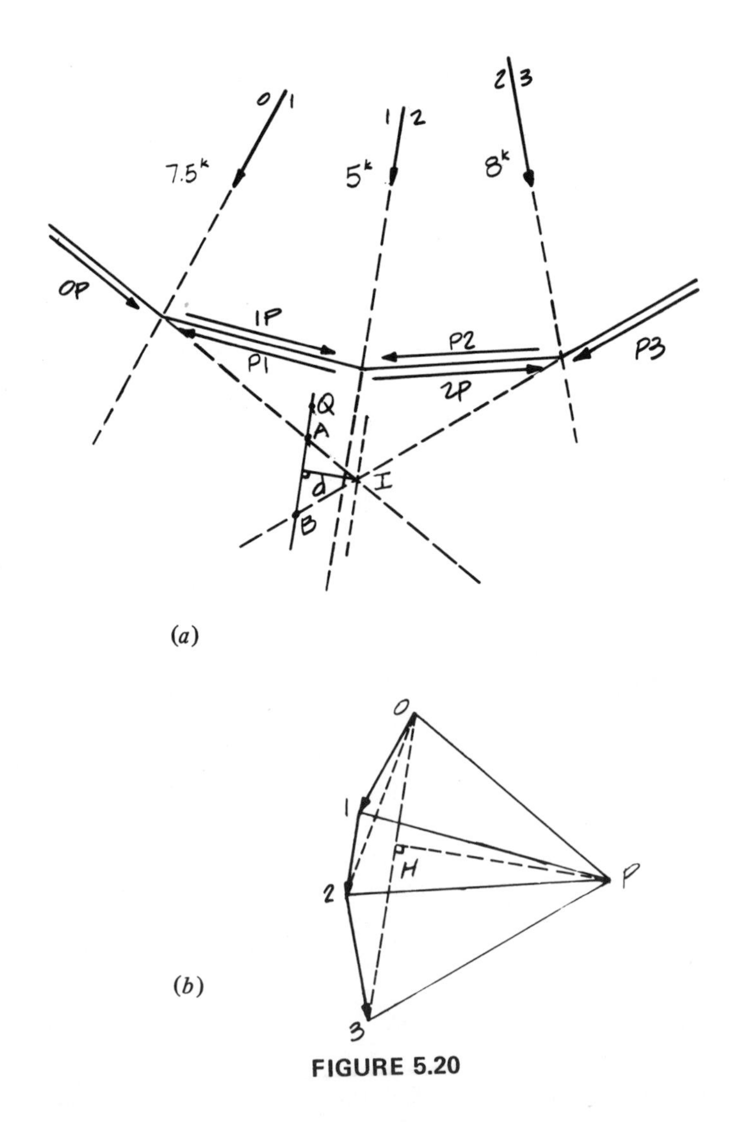

FIGURE 5.20

the force polygon are projected from an arbitrary point P called a pole.

To the left of the first force 0–1 in Figure 5.20a, we draw an arbitrary line parallel to segment $0P$ in Figure 5.20b. From the intersection point of this line with force 0–1, we now draw a new line parallel to $1P$. From the intersection of this line with force 1–2, we draw a parallel to $2P$, and so on. The polygon obtained in this manner is called a *string polygon* associated with the forces 0–1, 1–2, and 2–3 of the system, and with pole P.

If we extend the first and last sides of the string polygon, their intersection point I is on the line of action of the resultant of the system. The resultant itself is the dotted force 0–3 in Figure 5.20*b*, as it can be seen by combining by the triangle rule the dotted resultant 0–2 of forces 0–1 and 1–2 with the last force 1–3. Since 0–3 gives magnitude, direction, and sign of the resultant and point I defines its correct line of action, the resultant is completely identified.

To prove that point I belongs to the correct line of action of the resultant, we can use the triangle rule to replace force 0–1, with two components of magnitude 0–P and P–1 (Figure 5.20*b*), whose true lines of action are the first and second sides of the string polygon (Figure 5.20*a*). Similarly, we replace force 1–2 of the system with two components of magnitude 1–P and P–2 (Figure 5.20*b*) on the second and third sides of the string polygon (Figure 5.20*a*). Finally, we replace the last force of the system with two components of magnitude 2–P and P–3 (Figure 5.20*b*) along the third and fourth sides of the string polygon (Figure 5.20*a*).

Forces P–1 and 1–P on the second side of the string polygon cancel out one another; the same is true of forces P–2 and 2–P on the third side of the polygon. The given system is then reduced to the two forces 0–P and P–3 along the first and last sides, respectively, of the string polygon. Their resultant, which is also the resultant of the given system, must contain intersection point I.

The moment M of the given system of forces around an arbitrary point Q is the same as the moment of the resultant 0–3 about Q. This moment is obtained graphically as follows: A line parallel to the resultant is drawn through point Q (Figure 5.20*a*). The first side of the string polygon is extended as far as its intersection A with the preceding line. The last side of the string polygon is extended as far as its intersection B with the same line. The segment A–B is the graphic representation of the moment M.

To prove this statement and obtain a scale for moment M, we reason as follows: Triangle ABI in Figure 5.20*a* and triangle $03P$ in Figure 5.20*b* are similar, for AB is parallel to 03, AI is parallel to $0P$, and BI is parallel to $3P$. It is therefore true that

$$\frac{AB}{0\text{–}3} = \frac{d}{PH},$$

where d is the height of triangle ABI on base AB, and PH is the height of triangle $03P$ on base 0–3. Letting the scale of the force be

$$S_F = 8 \text{ k/in.}$$

and the scale of the drawing

$$S_D = 4 \text{ ft/in.},$$

the preceding equation can be rewritten as

$$AB\, S_F\, S_D\, PH = (0\text{–}3)\, S_F\, S_D\, d.$$

Since (0–3) S_F is the true resultant of the force system and $S_D\, d$ is the true arm of the resultant around Q, (0–3), $S_D S_F d$ as well as AB ($S_F\, S_D\, PH$) are the true moment M, of which AB is the graphic representation and $PHS_F\, S_D$ is the scale. In our case, PH measures 1.75 in., and AB measures 0.625 in.; hence, the moment scale is

$$PHS_F\, S_D = (1.75 \text{ in.})(8 \text{ k/in.})(4 \text{ ft/in.}) = 56 \text{ k-ft/in.},$$

and the moment is

$$M = (0.625 \text{ in.})(56 \text{ k-ft/in.}) = 35 \text{ k-ft.}$$

If last point N of the N-sided force polygon of a system of N forces coincides with its first point 0, then the resultant $0N$ of the system has zero magnitude (Figure 5.21*b*). It can, therefore, be said that

a closed force polygon is the graphic condition for force equilibrium.

In this case, projections $0P$ and $4P$ coincide, and the first side of the string polygon is parallel to the last side (Figure 5.19*a*).

The force system is equivalent to a *couple* of equal and opposite forces of magnitude $NP = 0P$ applied along the first and last sides of the string polygon. The moment of a couple around any point has constant value. Figure 5.21*a* shows, in fact, that segment AB cut by the first and last sides of the string polygon on any vertical line is the same. If the string polygon has its first and last sides not only parallel

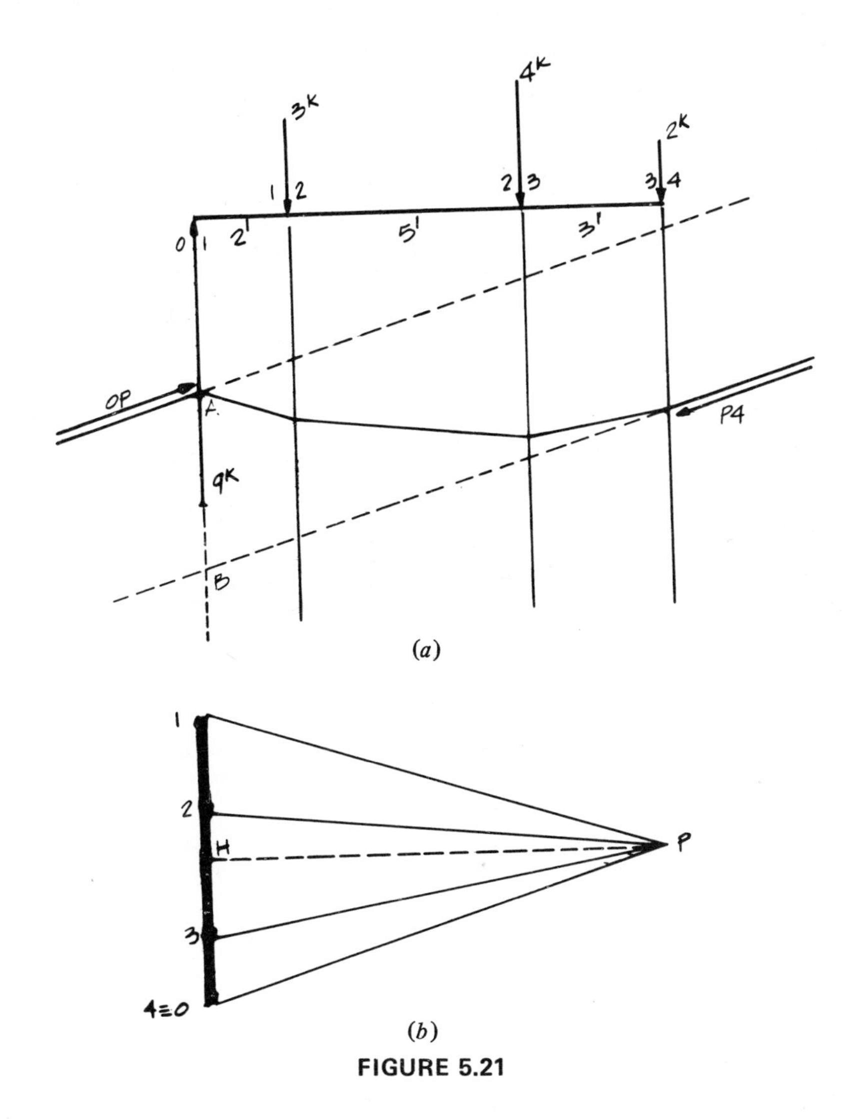

(a)

(b)

FIGURE 5.21

but also colinear, points A and B will coincide in every case (Figure 5.22a). The system of forces therefore has a moment AB equal to zero around every point in space. It can be said that

colinearity of the first and last sides of a string polygon is the graphic condition for moment equilibrium.

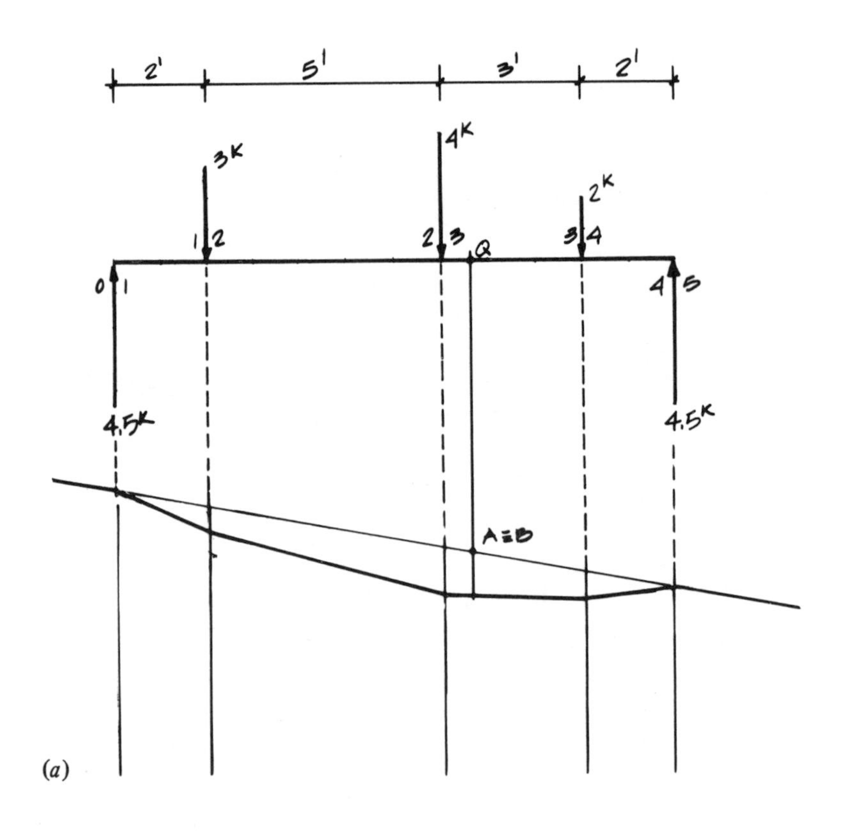

(a)

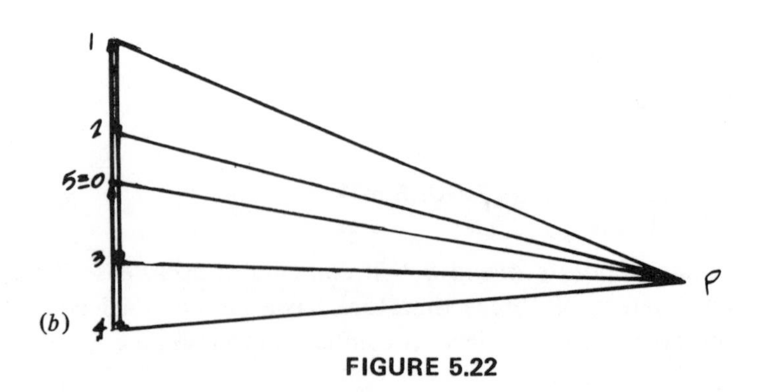

(b)

FIGURE 5.22

PROBLEMS

5.1. Replace the distributed loads with their resultants. Calculate reactions of the constraints, and draw diagrams of shear force and bending moment (Figures 5.23*a*, *b*, *c dotted*, and *d dotted*).

Solution. We replace the distributed loads with their resultants. Then

$$g(x) = \frac{-dV}{dx} = 0;$$

thus, the shear is constant between the concentrated external forces, and the moment varies linearly, since its derivative is constant

$$V(x) = \frac{dM}{dx} = \text{constant}.$$

The distributed load on the span l_1 is replaced by a concentrated force equal to the triangular area

$$\frac{l_1}{2} g_{max} = \frac{3}{2}(2) = 3 \text{ k}.$$

This force is applied at the center of the triangle, which is $2l_1/3 =$ 2 ft away from the tip of the overhang. Similarly, the distributed load on the span l_2 is replaced by a concentrated force equal to

$$\frac{l_2}{2} g_{max} = 9 \text{ k},$$

placed 6 ft away from the tip of the overhang. The resultant of the parabolic load on half of the central span has the magnitude (Problem 9.1)

$$\frac{1}{3} g_{max} \frac{l}{2} = 10 \text{ k},$$

and it is placed at the distance from the support (Problem 9.3)

$$\frac{1}{4}\left(\frac{l}{2}\right) = 3.75 \text{ ft}.$$

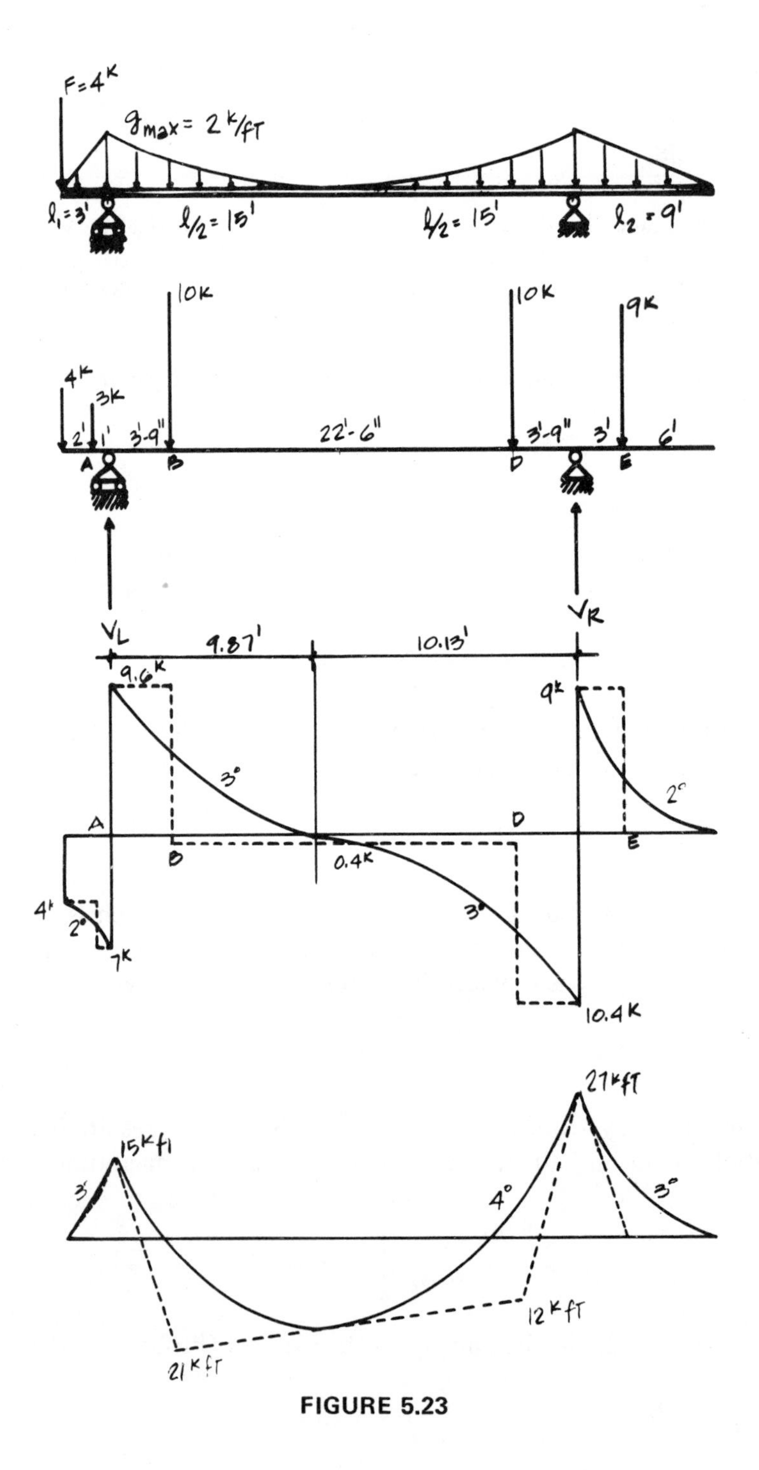
F=4K
g_{max}= 2 K/fT
l_1=3'
l/2= 15'
l/2= 15'
l_2=9'
10K
10K
9K
4K
3K
2'
1'
3'-9"
22'-6"
3'-9"
3'
6'
A
B
D
E
VL
VR
9.87'
10.13'
9.6K
9K
3°
2°
A
D
B
0.4K
E
4K
2°
7K
3°
10.4K
21KfT
15Kft
3°
4°
3°
12KfT
21KfT

FIGURE 5.23

The equation of equilibrium of moments around point A with the entire load of span l concentrated at midspan is

$$\Sigma M = 0 = l_1 F + \frac{l_1}{2} g_{\max} \left(\frac{l_1}{3}\right) - \frac{l}{3} g_{\max} \left(\frac{l}{2}\right) + l V_R - \frac{l_2}{2} g_{\max} \left(l + \frac{l_2}{3}\right)$$

Numerically,

$$0 = 3(4) + 1.5(2)1 - 10(2)15 + 30 V_R - 4.5(2)33,$$

from which

$$V_R = (-12 - 3 + 300 + 297)/30 = 19.4 \text{ k}.$$

The equation of vertical equilibrium yields

$$V_L = F + \frac{l_1}{2} g_{\max} + \frac{l}{3} g_{\max} + \frac{l_2}{2} g_{\max} - V_R .$$

Numerically,

$$V_L = 4 + 3 + 20 + 9 - 19.4 = 16.6 \text{ k}.$$

The shear diagram can now be drawn. The values shown in Figure 4.25*c dotted* are obtained as follows:

At the left end, $V = -4$ k.
At point A, $V = -4 - 3 = -7$ k.
At the roller, $V = -7 + 16.6 = 9.6$ k.
At point B, $V = 9.6 - 10 = -0.4$ k.
At point D, $V = -0.4 - 10 = -10.4$ k.
At the hinge, $V = -10.4 + 19.4 = 9$ k.
At the right end, $V = 9 - 9 = 0$ k.

The values of the bending moment shown in Figure 5.23*d dotted* are obtained as follows:

At the left end, $M = 0$.
At point A, $M = 4(2) = 8$ k-ft (same as shear area from left end to A).
At the roller, $M = 4(3) + 3(1) = 15$ k-ft (same as shear area on overhang).

At point B, $M = 4(6.75) + 3(4.75) - 16.6(3.75) = 21$ k-ft (same as shear area from left end to B).

Also at point B, $M = 15 - 9.6(3.75) = 21$ k-ft (moment at boundary of span algebraically added to shear area from roller to point B).

At point D, $M = 4(29.25) + 3(27.25) - 16.6(26.25) + 10(22.5) = -12$ k-ft.

Also at point D, $M = 15 - 9.6(3.75) + 0.4(22.5) = -21 + 9 = -12$ k-ft.

At the hinge, $M = 9(3) = 27$ k-ft (same as shear area on overhang).

5.2. Draw the correct curves that show the variation of shear force and bending moment along the beam axis (Figures 5.23*a–d*).

Solution. At the left end of the beam, the shear coincides with F. Along the span l_1 of the overhang, load g increases linearly (first-degree curve) from $g = 0$ to $g = g_{max}$. Therefore, the shear curve varies parabolically (second-degree curve) with a slope increasing from zero (horizontal tangent) at the tip to g_{max} at the support. At the left support, the shear curve has a step due to the concentrated force V_L.

Along the central span l, the load varies parabolically; thus, the shear varies with a third-degree curve. The slope of this curve decreases from g_{max} at the support to zero (horizontal tangent) at midspan, where it turns from decreasing to increasing (inflexion point). At the right support, a step in the shear curve is due to V_R. On the right overhang, the load is linear and the shear parabolic, with a slope decreasing from g_{max} to zero.

The bending moment is zero at the left end of the beam. The moment then varies with a third-degree curve. The slope of this curve (shear) increases from the tip to the support. At the support, the slope abruptly changes sign and value (cuspid or kink). Along the central span, the moment varies with a fourth-degree curve. The slope of this curve decreases as far as the point where the shear is zero (horizontal tangent, maximum moment on span). From there to the hinge, the slope of the moment curve increases with opposite sign. At the hinge, the slope abruptly changes sign and value. Along the right overhang, the moment varies with a third-degree curve. The slope of this curve vanishes at the tip (horizontal tangent).

Engineering students and those architecture students interested in the precision of mathematics can solve Problems 5.1 and 5.2 by following a systematic approach that is adaptable to every case: The load on span l varies according to a second-degree curve. Therefore, on span l,

$$g(x) = ax^2 + bx + c.$$

Placing the origin of the x-coordinate at the roller, coefficients a, b, and c of the curve equation are obtained by stating

At $x = 0$, $\quad g(x) = g_{max} = c.$

At $x = \frac{l}{2}$, $\quad g(x) = 0 = a\left(\frac{l}{2}\right)^2 + b\left(\frac{l}{2}\right) + g_{max}.$

At $x = l$, $\quad g(x) = g_{max} = a(l)^2 + b(l) + g_{max}.$

Multiplying both sides of the second equation by -4, then adding $-g_{max}$ to both sides of the third equation, we obtain

$$4g_{max} = -al^2 - 2bl,$$

$$0 = al^2 + bl.$$

Adding the two equations gives

$$4g_{max} = -bl,$$

from which we get

$$b = \frac{-4}{l} g_{max}.$$

Replacing b in the equation

$$0 = al^2 + bl,$$

we get

$$a = \frac{-b}{l} = \frac{4}{l^2} g_{max}.$$

Therefore,

$$g(x) = \left(\frac{4}{l^2} g_{max}\right) x^2 - \left(\frac{4}{l} g_{max}\right) x + g_{max},$$

with

$$g_{max} = 2 \text{ k-ft}$$

$$l = 30 \text{ ft}$$

$$g(x) = \frac{2}{225} x^2 - \frac{4}{15} x + 2.$$

The equation of the shear curve is

$$V(x) = -\int g(x)\, dx + V_0.$$

Replacing $g(x)$,

$$V(x) = -\frac{4}{l^2} g_{max} \int x^2\, dx + \frac{4}{l} g_{max} \int x\, dx - g_{max} \int dx + V_0.$$

Integrating

$$V(x) = -\left(\frac{4}{3l^2} g_{max}\right) x^3 + \left(\frac{2}{l} g_{max}\right) x^2 - (g_{max})\, x + V_0.$$

With the preceding values of l and g_{max}, with $V_0 = 9.6$ k, we obtain

$$V(x) = -\frac{2}{675} x^3 + \frac{2}{15} x^2 - 2x + 9.6.$$

The preceding expression of V allows us to plot the complete shear diagram on the span.

The maximum moment materializes at the x-coordinate where $V = 0$. This coordinate is therefore given by the equation

$$0 = -\frac{2}{675} x^3 + \frac{2}{15} x^2 - 2x + 9.6,$$

which, after multiplying both sides by $-675/2$, becomes

$$0 = x^3 - 45x^2 + 675x - 3240.$$

The solution of the equation is

$$x = 9.87 \text{ ft.}$$

The equation of the moment curve is

$$M(x) = \int V(x)\,dx + M_0.$$

Replacing $V(x)$,

$$M(x) = -\frac{4}{3l^2} g_{max} \int x^3\,dx + \frac{2}{l} g_{max} \int x^2\,dx - g_{max} \int x\,dx + V_0 \int dx + M_0.$$

Integrating

$$M(x) = -\left(\frac{1}{3l^2} g_{max}\right) x^4 + \left(\frac{2}{3l} g_{max}\right) x^3 - \left(\frac{1}{2} g_{max}\right) x^2 + V_0 x + M_0$$

with our values for l, g_{max}, V_0, and $M_0 = -15$ k-ft,

$$M(x) = -\frac{x^4}{1350} + \frac{2}{45}x^3 - x^2 + 9.6x - 15.$$

We realize from the moment curve (Figure 5.23*d*) that $\int V(x)\,dx$ and M_0 have opposite signs, since $M(x)$ decreases as x increases.

The preceding expression for M allows us to plot the complete diagram on span l. M_{max} can be found directly by replacing $x = 9.87$ ft in the expression $M(x)$.

$$M_{max} = -\frac{(9.87)^4}{1350} + \frac{2}{45}(9.87)^3 - (9.87)^2 + 9.6(9.87) - 15,$$

$$M_{max} = 18 \text{ k-ft.}$$

5.3. Replace the distributed loads with their resultants in order to draw approximate diagrams of shear and moment (Figures 5.24*a*, *b*, *c dotted*, and *d dotted*).

Solution. The linear load has a resultant

$$0.5(6)(1.5) = 4.5 \text{ k.}$$

This resultant is placed 4 ft away from point B. The constant load has a resultant

$$12(1.5) = 18 \text{ k}$$

placed 6 ft away from the roller. The parabolic load on the overhang has a resultant (see Problem 9.2)

$$(6)(1.5)\ 2/3 = 6 \text{ k}$$

placed $(3/8)(6) = 2.25$ ft away from the roller (Problem 9.4). The equation of vertical equilibrium is

$$\Sigma F_y = 0 = -5 - 4.5 - 18 + V_R - 6,$$

from which

$$V_R = 33.5 \text{ k}.$$

The equation of moment equilibrium around the right end is

$$\Sigma M = 0 = -M_L + 5(27) + 4.5(20) + 18(12) - 6(33.5) - 6(3.75),$$

from which

$$M_L = 262.5 \text{ k ft}.$$

The shear force has the following values on the dotted diagram in Figure 5.24:

At the slide,	$V = 0.$
At point A,	$V = -5$ k.
At point C,	$V = -5 - 4.5 = -9.5$ k.
At point E,	$V = -9.5 - 18 = -27.5$ k.
At the roller,	$V = -27.5 + 33.5 = 6$ k.
At point F,	$V = +6 - 6 = 0$ k.

The bending moment has the following values on the approximate dotted diagram in Figure 5.24*d*:

At the slide,	$M_L = -262.5$ k-ft constant as far as A for lack of shear force.
At point B,	$M = -262.5 + 5(3) = -247.5$ (same as algebraic

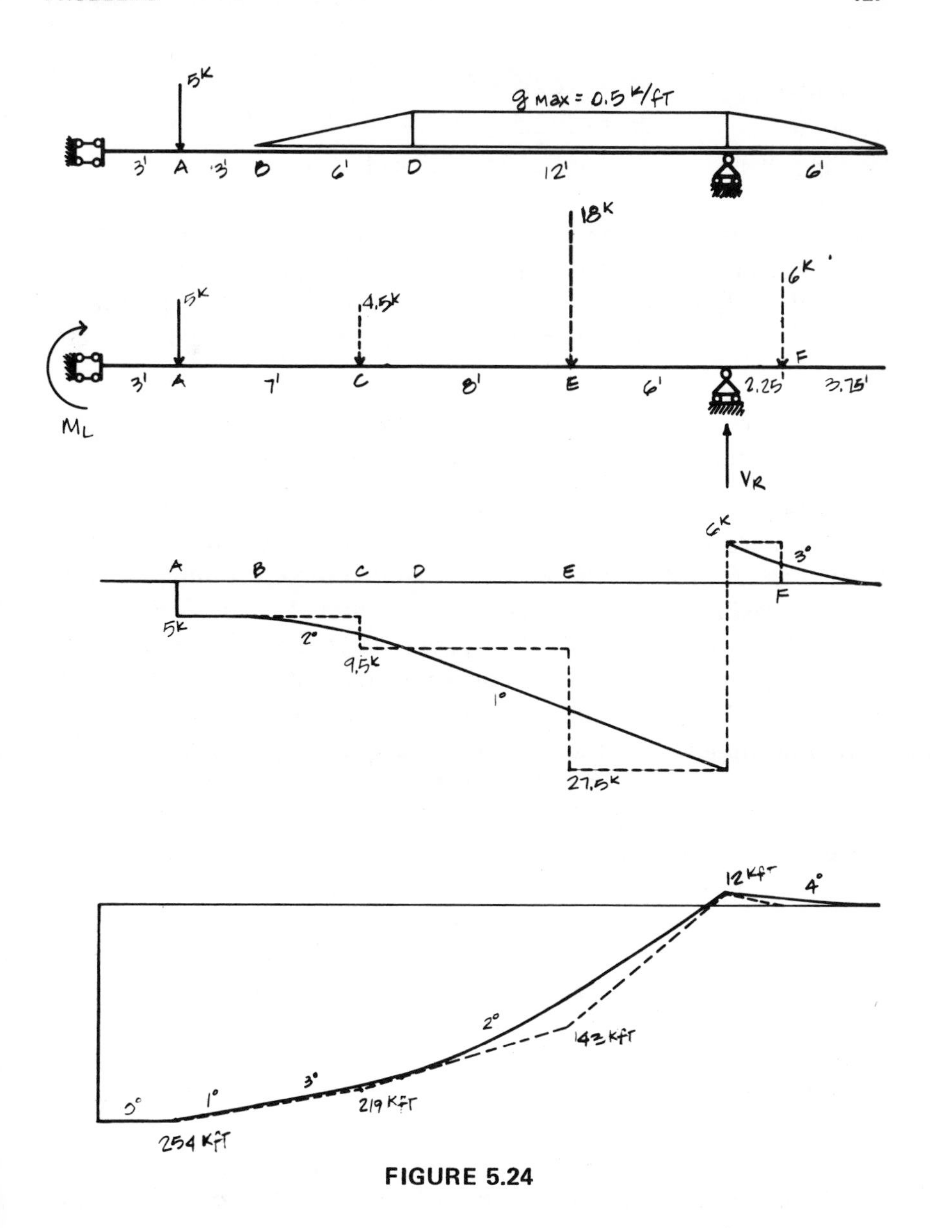

FIGURE 5.24

sum of boundary moment M_L with shear area from side to B).

At point C, $M = -262.5 + 5(7) = -227.5$ k-ft.

At point D, $M = -262.5 + 5(9) + 4.5(2) = -208.5$ k-ft.
At point E, $M = -262.5 + 5(15) + 4.5(8) = -151.5$ k-ft.
On the roller, $M = 6(2.25) = 13.5$ k-ft.

The values of the moment are correct from the slide to point B, at point D, and on the roller. The values at points C and E are approximate. Indeed, replacing the distributed loads with their resultants excludes the effect of some of the loads on the moment at these points.

5.4. Draw the correct curves that show the variation of shear and moment along the beam axis (Figures 5.24*a–c*).

Solution. At the left end of the beam, the shear vanishes because the slide does not react vertically. The shear remains constantly equal to zero as far as A, due to a lack of external forces. It plunges 5 k at A and remains constantly equal to 5 k from A to B. The load increases linearly from B to D. Thus, the shear varies parabolically with slopes increasing from zero (horizontal tangent) to 1.5. With this same slope, the shear varies linearly from point D to the roller.

At the support, a step in the shear curve is due to the reaction V_R. From the support to the right end, the load decreases parabolically. Then, the shear varies with a third-degree curve. The slope of this curve varies from 1.5 to zero (horizontal tangent).

M_L, the reactive moment of slide L, remains constant as far as point A, due to a lack of shear. The value of M_L is the largest on the span because the shear is zero at the left end. From A to B, where the shear is constant (zero-degree curve), the moment decreases linearly (first-degree curve) with slope 5. Starting with this slope, the moment varies with a third-degree curve from B to D and with a second-degree curve from D to the roller. The third- and second-degree curves have increasing slopes as far as the roller. Here, the slope of the moment curve changes sign and value (cuspid). On the overhang, the moment varies with a fourth-degree curve and slopes vanishing at the right end.

5.5. Perform the same tasks in Problem 5.1 for the beam in Figure 5.25.

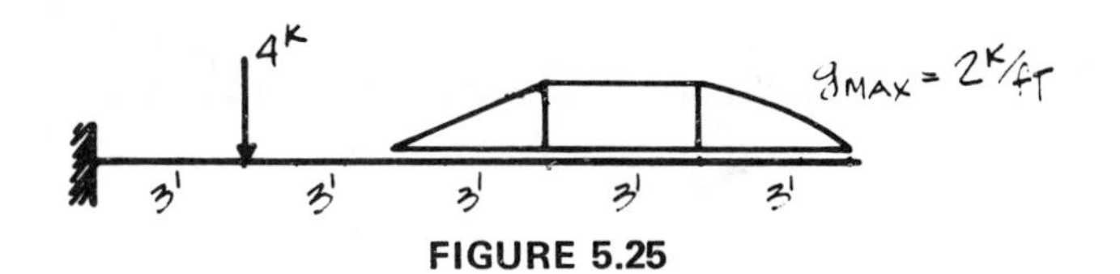

FIGURE 5.25

5.6. Perform the same tasks in Problem 5.2 for the beam in Figure 5.25.

5.7. Perform the tasks in Problem 5.1 for the beam in Figure 5.26. Divide the parabolic load into two equal parts.

5.8. Perform the tasks in Problem 5.2 for the beam in Figure 5.26.

5.9. Replace the distributed loads on the Gerber beam in Figure 5.27 with their resultants. Find the reactions of the constraints, and draw approximate diagrams of shear and moment.

Solution. The magnitude and position of the load resultants are shown in Figure 5.27*b*. Reactions of the constraints are evaluated by equations of equilibrium of the free bodies shown in Figure 5.27*c*. The solution must start with a free body on which only two reactions or interactions are unknown. Indeed, only two equations of equilibrium are available, $\Sigma F_y = 0$ and $\Sigma M = 0$.

Starting with the free body on the left, interaction V_c is obtained from the equation of moment equilibrium around point A

$$M = 0 = -4(8) + 8V_c,$$

from which

$$V_c = 4 \text{ k}.$$

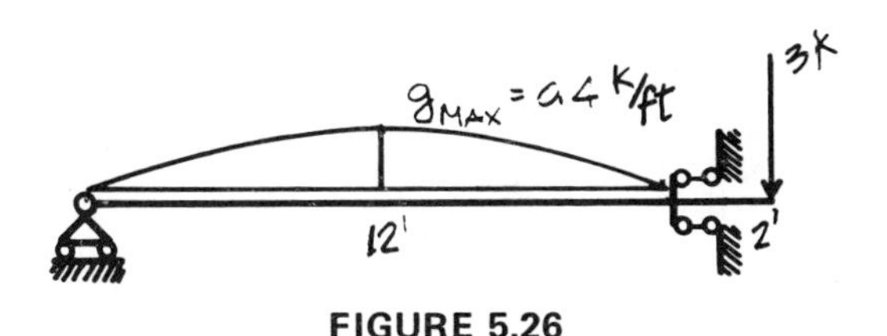

FIGURE 5.26

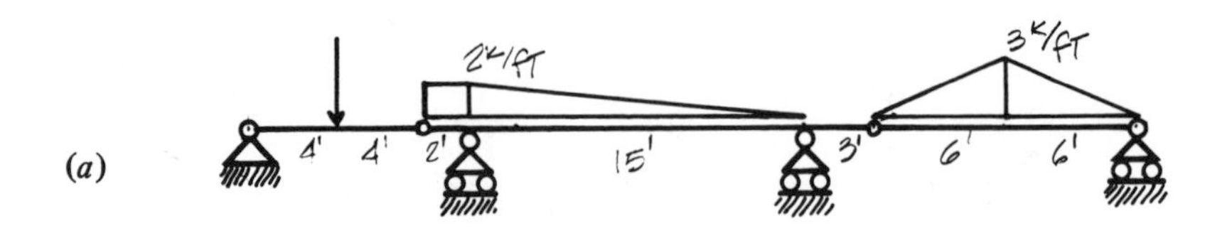
(a)
2K/FT
3K/FT
4'
4'
2'
15'
3'
6'
6'

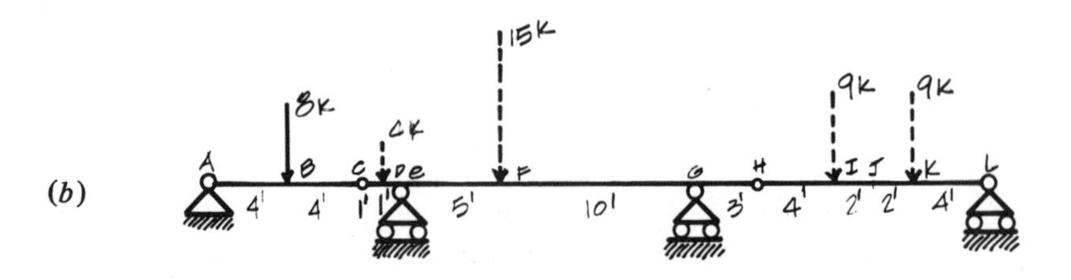
(b)
15K
8K
4K
9K
9K
A
B
C
D
E
F
G
H
I
J
K
L
4'
4'
1'
1'
5'
10'
3'
4'
2'
2'
4'

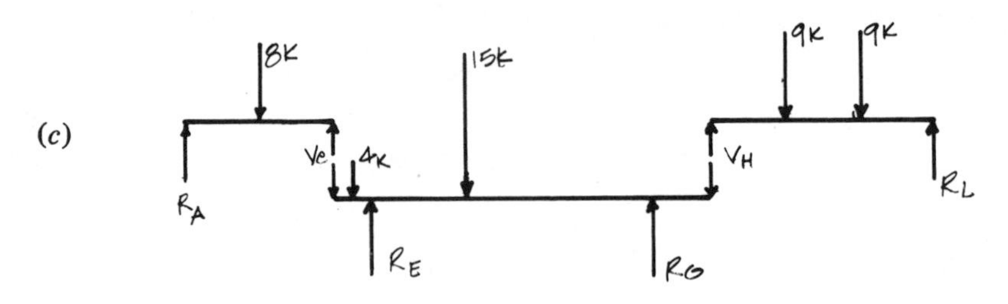
(c)
8K
15K
9K
9K
Vc
4K
VH
RA
RE
RG
RL

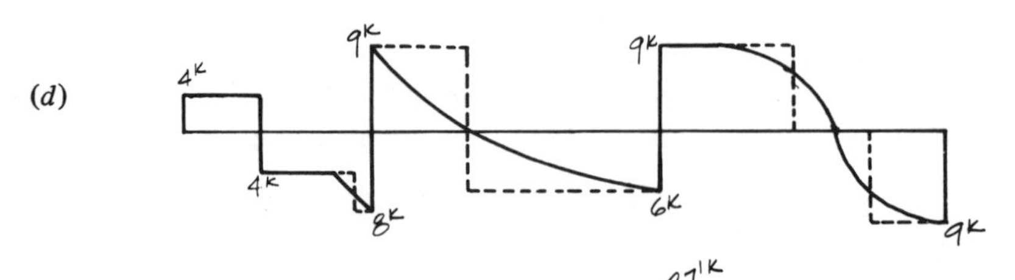
(d)
4K
9K
9K
4K
8K
6K
9K

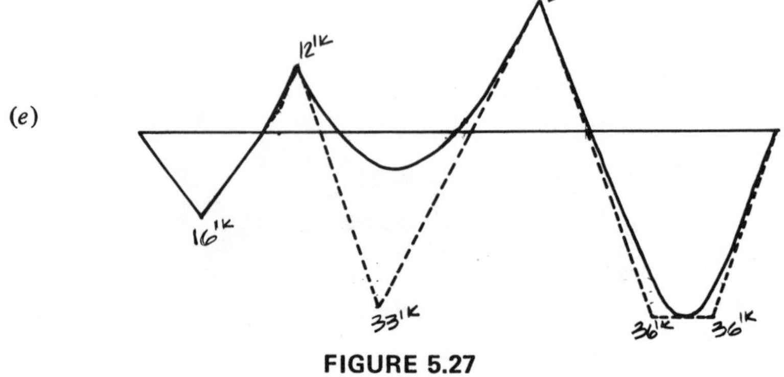
(e)
27'K
12'K
16'K
33'K
36'K
36'K

FIGURE 5.27

The equation of vertical equilibrium is

$$\Sigma F_y = 0 = R_A - 8 + 4,$$

from which

$$R_A = 4 \text{ k}.$$

Indeed, V_c and R_A could be evaluated by inspection because of symmetry.

The number of unknown reactions and interactions on the intermediate free body is reduced to three, which is still too high. However, the free body on the right has a manageable number of unknowns. Due to symmetry,

$$V_H = R_L = 9 \text{ k}.$$

Only R_E and R_G are now to be evaluated on the intermediate free body.

The equation of moment equilibrium around point E is

$$\Sigma M = 0 = 2V_C + 1(4) - 4(15) + 15R_G - 18V_H,$$

from which

$$R_G = (-8 - 4 + 75 + 162)/15 = 15 \text{ k}.$$

The equation of vertical equilibrium is

$$\Sigma F_y = 0 = -4 - 4 + R_E - 15 + 15 - 9,$$

from which

$$R_E = 17 \text{ k}.$$

The approximate shear diagram is shown by a dotted line in Figure 5.27*d*. At point A, the shear coincides with the 4-k reaction R_A. For lack of external forces, the shear remains constant as far as point B, where it plunges 8 k to a value of −4 k. At point D, the 4-k load changes the shear value from −4 to −8 k. The 17-k reaction R_E turns the shear from −8 to +9 k. At point F, the 15-k load makes the shear value plunge from 9 to −6 k. At point G, the 15-k reaction reverses the shear from −6 to +9 k. Due to a lack of external forces, this value of the shear remains unchanged as far as point I, where the shear vanishes due to the 9-k load. At point K, the shear plunges to −9 k and remains constant to the end L, where the 9-k reaction R_L makes the shear diagram return to its base line.

The approximate moment diagram is shown by a dotted line in Figure 5.27*e*. The values of the bending moment are calculated by progressive algebraic sums of the shear areas; thus,

At A, $M = 0$.
At B, $M = 4(4) = 16$ k-ft.
At C, $M = 16 - 4(4) = 0$ k-ft (hinge).
At D, $M = 0 - 4(1) = -4$ k-ft.
At E, $M = -4 - 8(1) = -12$ k-ft.
At F, $M = -12 + 9(5) = 33$ k-ft.
At G, $M = 33 - 6(10) = -27$ k-ft.
At H, $M = -27 + 9(3) = 0$ k-ft (hinge).
At I, $M = 0 + 9(4) = 36$ k-ft.
At K, $M = 36 + 0 = 36$ k-ft.
At L, $M = 36 - 9(4) = 0$ k-ft.

5.10. Following the reasoning in Problems 5.2 and 5.4, explain the correct shear diagram (solid line in Figure 5.27*d* of the beam shown in Figure 5.27*a*).

5.11. Following the reasoning in Problems 5.2 and 5.4, explain the correct moment diagram (solid line in Figure 5.27*e*) on the beam in Figure 5.27*a*.

5.12. Replace the distributed loads on the Gerber beam in Figure 5.28*a* with their resultants. Find the reactions of the constraints, and draw approximate diagrams of shear and moment (Figures 5.28*a–c*, *d dotted*, and *e dotted*).

Solution. The magnitude and position of the load resultants are shown in Figure 5.28*b*. The free bodies of the solid parts of the beam are shown in Figure 5.28*c*. The reactions of the constraints are obtained from equations of vertical force and moment equilibrium, applied to the free bodies of the beam's solid parts.

On all solid parts but EG, the number of unknown reactions exceeds two. The solution must, therefore, start with equations of

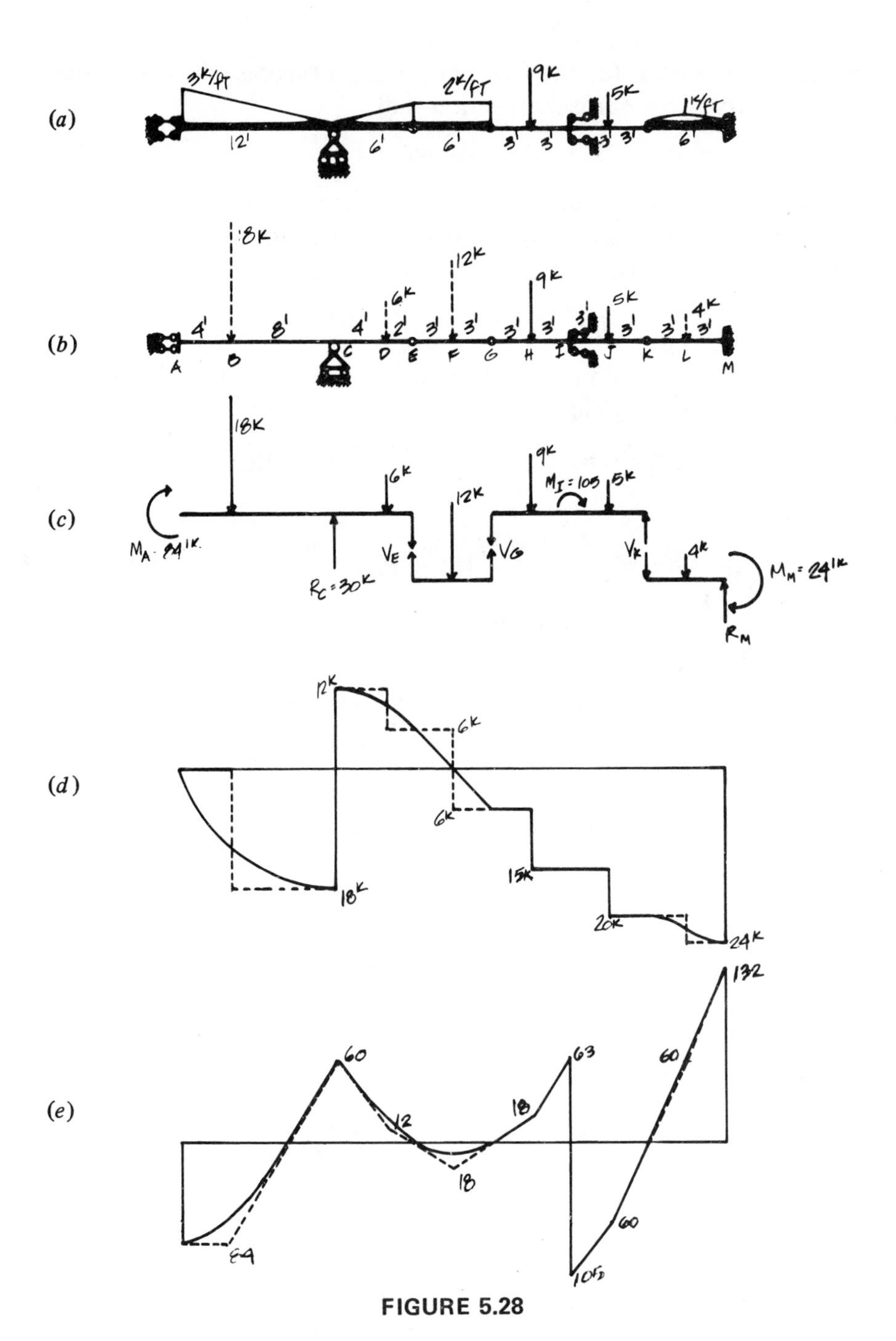
(a)
3K/FT
2K/FT
9K
5K
1K/FT
12'
6'
6'
3'
3'
3'
3'
6'
(b)
18K
12K
9K
6K
5K
4K
4'
8'
4'
2'
3'
3'
3'
3'
3'
3'
3'
3'
A
B
C
D
E
F
G
H
I
J
K
L
M
(c)
18K
9K
6K
12K
5K
M_I = 105
M_A = 84'K
V_E
V_G
V_K
4K
M_M = 24'K
R_C = 30K
R_M
(d)
12K
6K
6K
18K
15K
20K
24K
(e)
132
60
63
60
12
18
18
84
60
105

FIGURE 5.28

equilibrium written for the free body EG. In this case, however, due to symmetry, we see from inspection that

$$V_E = V_G = 6 \text{ k}.$$

There is now only one unknown vertical force R_C on AE. R_C must balance the algebraic sum of all other vertical forces or AE would move vertically; thus,

$$R_C = 18 + 6 + V_E = 30 \text{ k}.$$

An equation of moment equilibrium on AE pivoted, for example, around point E yields the value of M_A

$$\Sigma M = 0 = -M_A + 14(18) - 6(30) + 2(6)$$

from which

$$M_A = +252 - 180 + 12 = 84 \text{ k-ft}.$$

On GK, V_K is the only unknown vertical force, and by inspection, we see that

$$V_K = 6 + 9 + 5 = 20 \text{ k}.$$

An equation of moment equilibrium around pivot K (or any pivot) yields M_I

$$\Sigma M = 0 = 6(12) + 9(9) - M_I + 5(3).$$

Thus,

$$M_I = 72 + 81 + 15 = 168 \text{ k-ft}.$$

The value of R_M is also determined by inspection to be

$$R_M = 20 + 4 = 24 \text{ k},$$

and the equilibrium of moments on the solid part KM around pivot K requires that

$$\Sigma M = 0 = -4(3) + 6(24) - M_M,$$

from which

$$M_M = -12 + 144 = 132 \text{ k-ft}.$$

With all external forces and moments defined, the approximate diagram of shear forces (Figure 5.28*d dotted*) is drawn as follows: At point A, the shear vanishes (the slide does not react vertically). At point B, the shear plunges 18 k and remains constant as far as point C. At point C, the vertical reaction $R_C = 30$ k turns the shear from -18 to $+12$ k. This value remains constant as far as point D, where the 6-k load resultant reduces the shear to $+6$ k. From D to F, the shear keeps the value $+6$ k. The 12-k load resultant at point F turns the shear from $+6$ to -6 k, which remains unchanged as far as point H. At H, the shear takes a 9-k plunge and remains constantly equal to -15 k from H to J. The shear diagram, indeed, ignores the sliding constraint at point I, since the slide does not react with a vertical force. At J, the shear deepens to -20 k for the 5-k load. From J to L, the approximate shear diagram stays at the -20-k level. It plunges to -24 k at point L for the 4-k load resultant and remains constant to the end M, where $R_M = 24$ k takes the diagram back to its base line.

The approximate diagram of bending moments (Figure 5.28*e dotted*) is obtained from progressive algebraic sums of the shear areas under the dotted outline and the boundary moments. Thus, at point A, the moment is $M_A = 84$ k-ft (clockwise). In the approximate diagram, this moment remains constant as far as point B, due to a lack of shear. From point B, the 18-k load introduces counterclockwise moments to be subtracted from M_A. At C, the moment equals the algebraic sum of the boundary moment M_A, with the rectangular shear area extending from B to C.

$$M_C = -84 + 18(8) = +60 \text{ k-ft.}$$

By progressive algebraic sum,

$$M_D = 60 - 12(4) = 12 \text{ k-ft.}$$

$$M_E = 12 - 6(2) = 0 \text{ k-ft (hinge).}$$

$$M_F = -6(3) = -18 \text{ k-ft.}$$

$$M_G = -18 + 6(3) = 0 \text{ k-ft (hinge).}$$

$$M_H = 0 + 6(3) = 18 \text{ k-ft.}$$

$$M_I = 18 + 15(3) = 63 \text{ k-ft (left of } I\text{).}$$

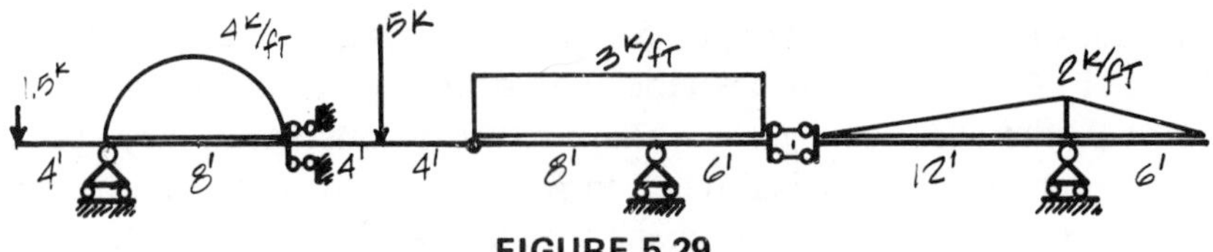

FIGURE 5.29

$M_I = 63 - 168 = -105$ k-ft (168 is a boundary moment).

$M_J = -105 + 15(3) = -60$ k-ft.

$M_K = -60 + 20(3) = 0$ k-ft (hinge).

$M_L = 0 + 20(3) = 60$ k-ft.

$M_M = 60 + 24(3) = 132$ k-ft $= M_R$.

5.13. Explain the degree, slope, inflection points, cuspids, or lack of them for the correct shear curves (Figure 5.28*d solid*).

5.14. Perform the same tasks in Problem 5.13 for the correct moment diagram (Figure 5.28*e solid*).

5.15. Perform the same tasks in Problem 5.12 for the Gerber beam in Figure 5.29.

5.16. Correct the approximate shear diagram drawn in Problem 5.15 with the exact shear diagram.

5.17. Correct the approximate moment diagram drawn in Problem 5.15 with the exact moment diagram.

5.18. Perform the same tasks in Problem 5.12 for the Gerber beam in Figure 5.30.

5.19. Correct the approximate shear diagram drawn in Problem 5.18 with the exact shear diagram.

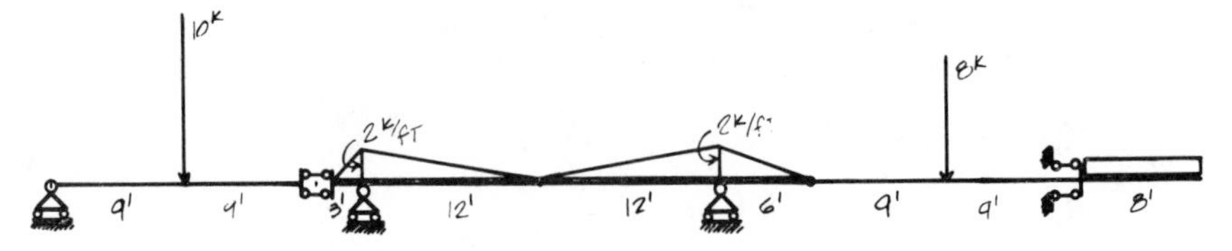

FIGURE 5.30

5.20. Correct the approximate moment diagram drawn in Problem 5.18 with the exact moment diagram.

5.21. Replace the distributed loads with their resultants. Using the graphic approach, evaluate the reactions of the constraints, and draw the approximate moment and shear diagrams (dotted diagrams; Figure 5.31).

Solution. Using the hinged section as the origin of the x-coordinates, the load resultants are given the following values and positions:

Resultant	1–2	2–3	3–4	4–5
Value	(2/3)(3)8 = 16 k See Problem 9.2.	16 k	6 k	12 k
x-coordinate	(5/8)(8) = 5 ft See Problem 9.4.	11 ft	20 ft	25 ft

The force polygon of the resultants is drawn using the scale of forces S_F = 10 k/in. An arbitrary pole P is placed at a perpendicular distance PH = 3 in. from the resultant 1–5 of the loads. A string polygon associated with the loads and pole P is drawn. Its sides, in order, are parallel to 1P, 2P, 3P, 4P, and 5P. This string polygon is the approximate moment curve, and the base line of the moment diagram is the closing side of the string polygon 0P = 6P. Approximate values of the moments shown in Figure 5.31 are read from this diagram with the scale

$$S_M = PHS_F S_D = (3 \text{ in.})(10 \text{ k/in.})(5 \text{ ft/in.}) = 150 \text{ k-ft/in.}$$

To draw the moment diagram again with a horizontal base line, a new pole P' is placed on the horizontal line that divides V_L from V_R on the force polygon. The polar distance $P'0$ must be the same as PH = 3 in. if S_M cannot change. The new string polygon with sides parallel to 0P', 1P', 2P', 3P', 4P', 5P', and 6P' is the desired moment diagram with horizontal base line. The reactions of constraints V_L = 0–1 and V_R = 5–6 are obtained by breaking the equilibrant 5–1 of the loads with a line P0 parallel to the first and last sides of the string polygon.

The base line of the shear diagram is drawn horizontally from

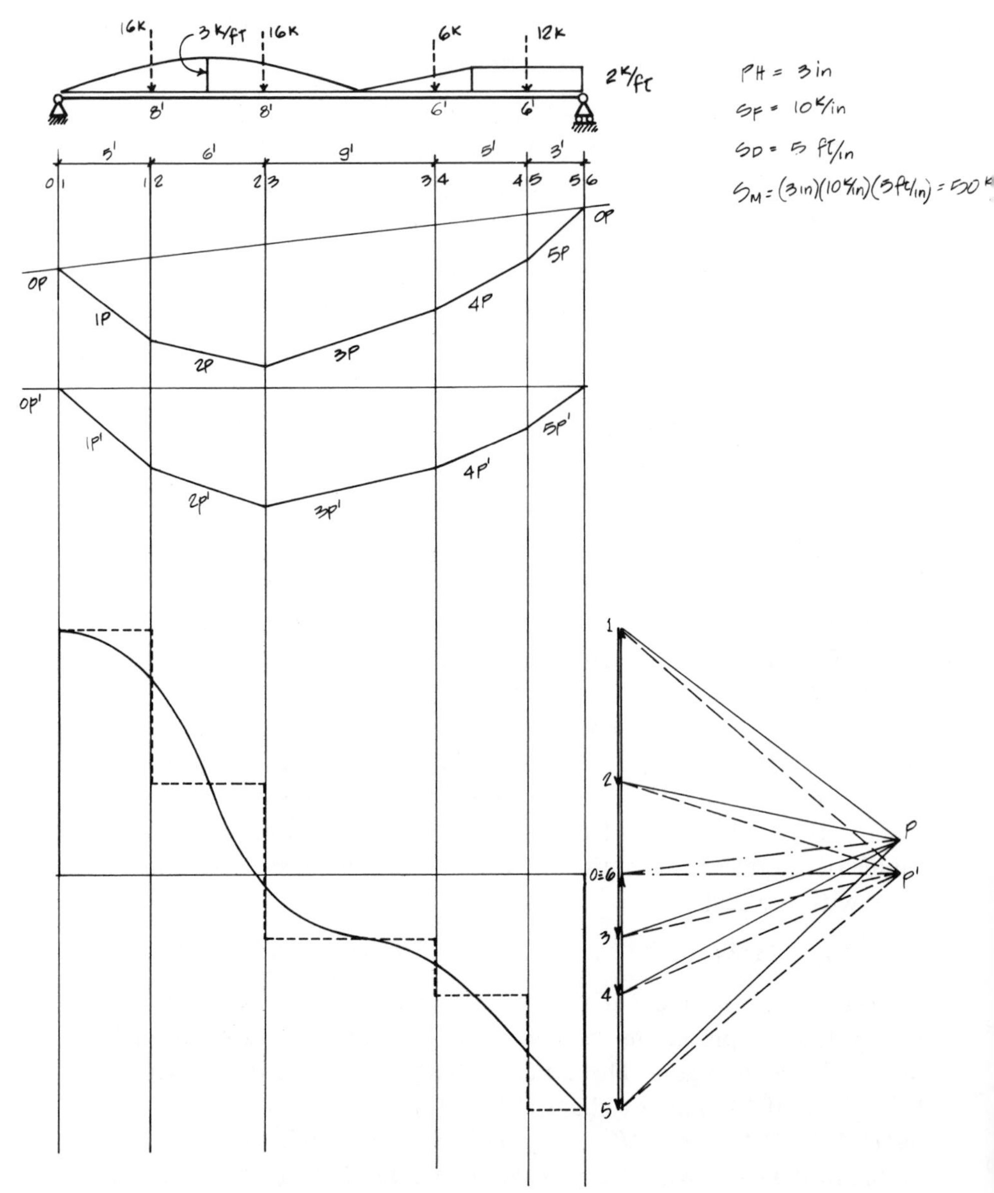

FIGURE 5.31. (*Original drawing reduced by 50%.*)

point 0 = 6 of the force polygon. The approximate shear diagram is obtained by horizontally projecting loads and reactions from the force polygon onto their actual lines of action. The values of the shear forces shown in Figure 5.31 are read from the scale

$$S_F = 10 \text{ k/in.}$$

5.22. Explain degree, slope, inflextion points, and cuspids, or lack of them, of the exact shear curves and moment curves (solid diagrams; Figure 5.31).

5.23. Using the graphic approach, draw the moment diagram for the beam in Figure 5.24*b*, and find the reactions of the constraints (Figure 5.32).

Solution. The force polygon of the loads is obtained by drawing them with the scale S_F = 10 k/in. in the order in which they are read from left to right. An arbitrary pole P is placed at a perpendicular distance PH = 2 in. from the resultant of the loads. Then, the string polygon associated with the loads and pole P is drawn with sides parallel to $1P$, $2P$, $3P$, $4P$, and $5P$.

The beam in this problem has two spans, one of which is the 6-ft overhanging span. In the case of multi-span beams, the moment diagram is obtained by associating the string polygon of the loads with as many base lines as there are spans. Two base lines must, therefore, be drawn for the beam in this problem.

The base line of each span is drawn according to two boundary conditions (specified moment values) available on the span. This is, in graphics, the equivalent task of numerically evaluating the values of the two constants of integration C_1 and C_2 in the analytic expression of the moment

$$M(x) = C_2 + \int C_1\, dx + \int dx \int g(x)\, dx.$$

Examining the beam in Figure 5.32, we see that the left span has only one boundary condition, which is that the moments are constant from the slide as far as load 0–1, since the shear

$$V = \frac{dM}{dx}$$

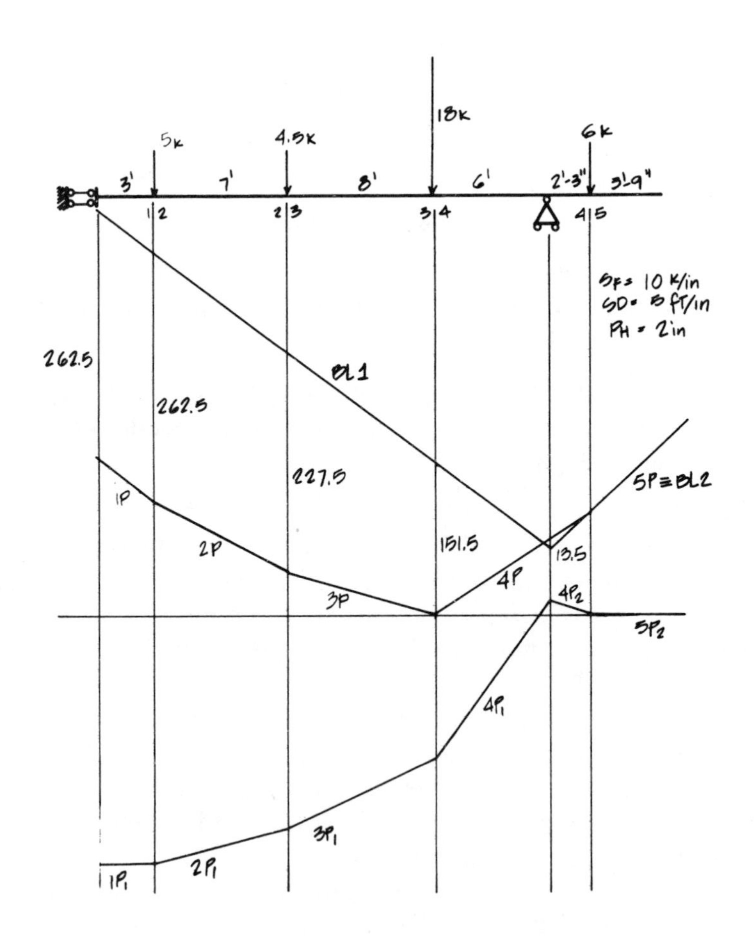

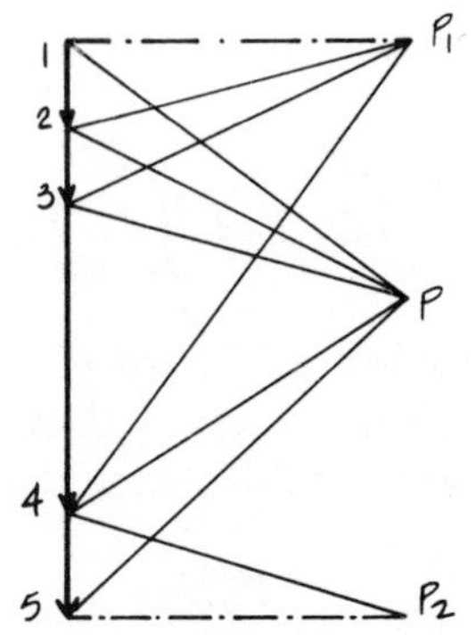

FIGURE 5.32. (*Original drawing reduced by 50%.*)

vanishes on that part of the beam. This suggests that base line $BL1$ of the left span must be parallel to side $1P$ of the string polygon; otherwise, the moment would not be constant. More information is needed, however, in order to correctly place the base line, and at this point it is not available.

We therefore shift our attention to the other span, where a sufficient number of boundary conditions exists. Indeed, the bending moment must vanish everywhere on the terminal unloaded arm of the overhang. The only base line that meets this specification is side $5P = BL2$ of the string polygon. This base line must extend as far as the span, which means as far as the roller. The value of the moment on the roller is now defined, and it provides the second boundary condition for the left span, since the value of the moment on the roller is the same for both spans.

With two boundary conditions now available, the base line of the left span is uniquely defined. It coincides with the parallel to side $1P$ of the string polygon that makes the moment on the left of the roller equal to the moment on the right. The values of the moments shown in Figure 5.32 are read with the scale $S_M = 100$ k-ft/in.

To obtain the moment diagram with the same horizontal base line for both spans, we first draw a parallel $P1$ to $BL1$ from pole P. A new pole P_1 is placed on a horizontal line drawn from point 1. The distance $1P_1$ is the same as $PH = 2$ in. A string polygon associated with the forces on the left span and pole P_1 is drawn. Next, we draw a parallel $P5$ to $BL2$ from P. Pole P_2 is placed on a horizontal line drawn from point 5. The distance $5P_2$ is the same as $PH = 2$ in. A string polygon associated with the forces on the overhang and pole P_2 is drawn. This polygon must continue the polygon drawn with pole P_1, thereby completing the moment diagram with a horizontal base line common to both spans.

The reactions of the constraints are found as follows. Line $P1$ drawn from pole P and parallel to base line $BL1$ breaks the equilibrant 4–1 of the load on the span at point 1, between shear 1–1 at the left end and shear 4–1 at the right end of the span. Line $P5$ drawn from pole P and parallel to base line $BL2$ breaks the equilibrant 5–4 of the load on the overhang at point 5, between shear 5–5 at the tip and shear 5–4 at the left end of the overhang. Shear 1–1 (equal to zero) at the left end of the span is the reaction of the slide.

Shear 4–1 at the right end of the span added to shear 5–6 at the left end of the overhang gives reaction 5–1 of the roller. Shear 5–5 (equal to zero) at the right end of the overhang is the reaction of the free end.

5.24. Replace the distributed loads with their resultants, and perform the same tasks as in Problem 5.23 (Figure 5.26).

5.25. Replace the distributed loads with their resultants and perform the same tasks as in Problem 5.23 (Figure 5.23).

5.26. Replace the distributed loads with their resultants and perform the same tasks as in Problem 5.23 (Figure 5.25).

5.27. Perform the same tasks in Problem 5.23 for the Gerber beam in Figure 5.28*b* (Figure 5.33).

Solution. The force polygon and the string polygon associated with the load and pole P are drawn as usual. To obtain the moment diagram, the string polygon must be associated with as many base lines as there are spans. In this case, the Gerber beam has three spans: AC, CI, and IM. Thus, three base lines–$BL1$, $BL2$, and $BL3$–are needed to complete the moment diagram. The only specification on span AC is that the moment must be constant from A to B where the shear vanishes. A second boundary condition is not readily available. Similarly, the only usuable specification on span IM is that the moment must vanish at hinge K. A second specification is lacking. On the intermediate span CI, however, two boundary conditions are available: The moment must vanish at points E and G. Thus, base line $BL2$ must cross side $2P$ of the string polygon at hinge E and side $3P$ of the string polygon at hinge G. Moment M_C on roller C is now defined, and it provides the second boundary condition needed on span AC in order to draw its base line $BL1$.

Because slide I does not react with a vertical force, the shear (slope of the moment curve) has the same value on both sides of I. This implies that side $4P$ of the moment diagram is inclined with the same angle on both base lines $BL2$ and $BL3$. Thus, $BL2$ and $BL3$ are parallel. $BL3$ is, then, the parallel to $BL2$ that crosses $5P$ at hinge K where the moment must vanish.

In order to obtain a moment diagram with a single horizontal base line, we draw the parallel PO to base line $BL1$ from pole P. A new

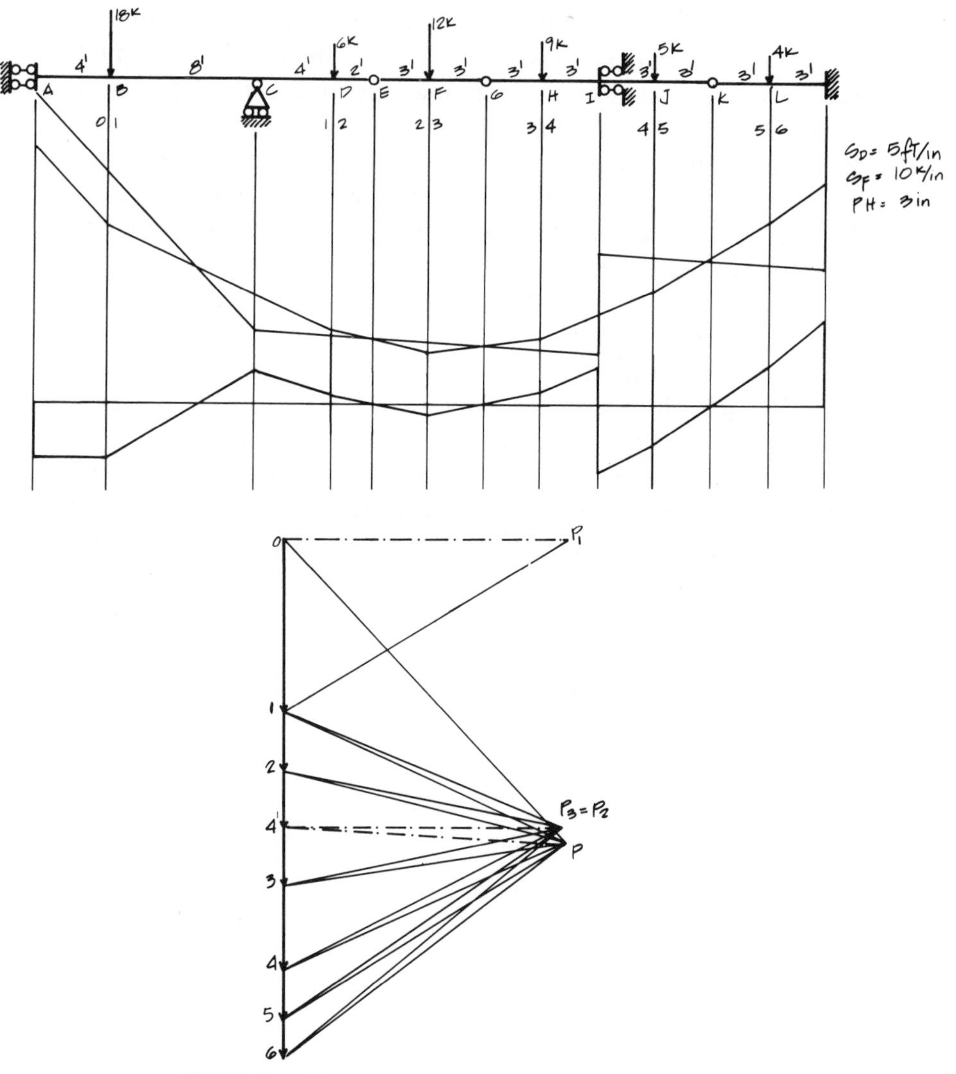

FIGURE 5.33. (*Original drawing reduced by 50%.*)

pole P_1 is placed on the horizontal line OP_1 drawn from point O. The distance OP_1 is the same as the polar distance $PH = 3$ k. A string polygon associated with the forces on span AC is drawn using pole P_1.

Next, we draw a line parallel to $BL2$ from P. From point 4′ where this line crosses the equilibrant 4–1 of the load on CI, we draw a horizontal line and place a pole P_2 3 in. away from 4–1 on this line. A string polygon associated with pole P_2 and the forces on span CI is drawn as a countinuation of the string polygon of span AC. Finally, we draw a line parallel to $BL3$ from P. From point $4'$ where this line crosses the equilibrant 6–4 of the load on span IM, we draw a horizontal line and place a pole P_3 3 in. away from 6–4 on this line.

Since $BL2$ and $BL3$ are parallel, pole P_2 coincides with pole P_3. A string polygon associated with the forces on span IM and pole P_3 is drawn under the condition that its side $5P_3$ crosses the horizontal base line common to all three spans at the coordinate of hinge K.

The reactions of the constraints are found as follows: Parallel $P0$ of base line $BL1$ breaks the equilibrant 1–0 of the load on span AC at point 0, between the left-end shear V_A = 0–0 and shear V_{CA} = 1–0 at the right end of span AC. Parallel $P4'$ of base line $BL2$ from pole P breaks the equilibrant 4–1 of the load on span CI between the span's left-end shear V_{CI} = 4′–1 and the right-end shear V_{IC} = 4–4′. Finally, parallel $P4'$ of base line $BL3$ drawn from pole P breaks the equilibrant 6–4 of the load on span IM at point $4'$, between the left-end shear 4′–4 and the right-end shear 6–4′.

Reaction R_A coincides with V_A = 0–0. Thus,

$$V_A = 0 \text{ k (slide)}.$$

Reaction R_C is given by the sum

$$R_C = V_{CI} + V_{CA} = 4'\text{–}1 + 1\text{–}0 = 4'\text{–}0.$$

Reading 4′–0 from the scale S_F = 10 k/in., we obtain

$$R_C = 30 \text{ k}.$$

Reaction R_I (slides do not react vertically) is given by the sum

$$R_I = V_{IC} + V_{IM} = (4\text{–}4') - (4'\text{–}4).$$

Thus,

$$R_I = 0 \text{ k}.$$

Finally, reaction R_M is given by

$$R_M = V_{MI} = 6\text{–}4'.$$

From the scale $S_F = 10$ k/in., we read

$$R_M = 24 \text{ k.}$$

5.28. Perform the same tasks in Problem 5.23 for the Gerber beam in Figure 5.27.

5.29. Perform the same tasks in Problem 5.23 for the Gerber beam in Figure 5.28.

5.30. Perform the same tasks in Problem 5.23 for the Gerber beam in Figure 5.29.

5.31. Perform the same tasks in Problem 5.23 for the Gerber beam in Figure 5.30.

The terminal building at the John Foster Dulles International Airport in Virginia near Washington, D.C. Architect: Eero Saarinen. Structural engineers: Ammann and Whitney, New York. Photograph by Ezra Stoller/ESTO.

SIX

CABLES

It was said in introducing the study of internal forces that most structures are subject to a combination of a few and occasionally all the internal forces. Joists, beams, and girders, for example, are often called flexural members because bending moment, and its derivative, the shear, are produced in these horizontal elements by vertical loads. Axial forces are produced by vertical loads in nonhorizontal beams or by nonvertical loads in horizontal beams. Twisting moments are induced by vertical loads external to the plane of the beam axis (balcony slabs of spandrel beams) or placed on beam axes not contained in vertical planes (beams curved in plane). Columns, arches, and bars of trusses are subject to axial forces, bending, and shear, except under ideal theoretic conditions. All four internal forces usually occur in two-dimensional and three-dimensional structures, such as plates, space grids, and shells.

Cables are exceptional because only pure tension is induced in them by loads of any extraction. The reason for this unique feature is to be found in a cable's ability to adapt the shape of its geometric axis to changing loads and in its inability to carry compressive forces. Anyone trying to exert a compressive force along a very thin string would immediately realize the impossibility of the task. Taking an accordion as an example of a structural element, we saw that bending produces shrinkage along the lower or upper edge of the side elevation, and shear and torsion produce shrinkage of one diagonal on the side elevation. It is thus evident that none of these internal forces can materialize in a structural element that will not carry compression.

The shape instability of cables and their lack of compressive strength depends on the extreme thinness of these structures. Spans hundreds of feet long can usually be negotiated by cables only a few inches in diameter, which is, in turn, explained by the original state-

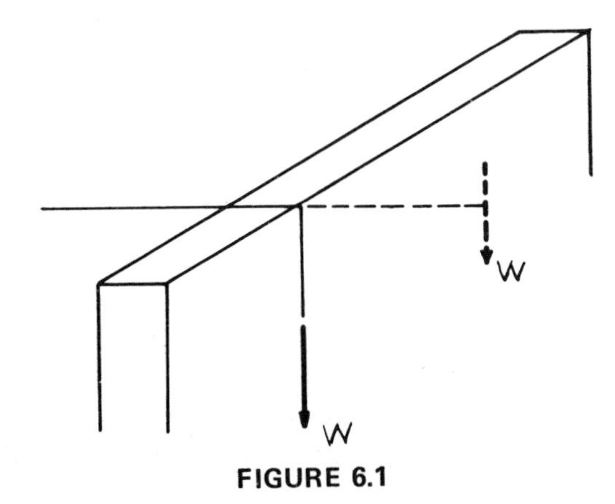

FIGURE 6.1

ment that only uniform tension, the most efficient of all stress patterns, exists on cross sections of a cable. To understand how a cable can change its shape and transfer loads to the supports in pure tension, we consider a span cantilevering from a wall with a vertical load at its free end (Figure 6.1).

A rigid structure such as a beam transfers the load to the supporting wall by a shear-bending mechanism, as explained in Chapter 4, Figure 4.1*b*. A thin cable, instead, will rotate rigidly to a vertical position on which the load and structural axis are colinear. Without lever arms around any of the points of the cable, the vertical force will not produce moments but only axial tension. Accordingly, the difference between cables and other structures can be identified with the difference between rigid motion practically without deformation and deformation without rigid motion. In terms of statics, the definitions of deformation and rigid motion are:

Deformations are geometric changes that occur when a system of loads or distortions (creep, shrinkage, thermal variations, foundation settlements) are applied to structures sufficiently constrained (statically determinate and redundant structures). Deformations are elastic if they vanish on removal of the loads and plastic if they are permanent.

Rigid motions are geometric changes that occur when loads are applied to structures that are externally or internally unstable.

For example, an open door has three degrees of freedom and two

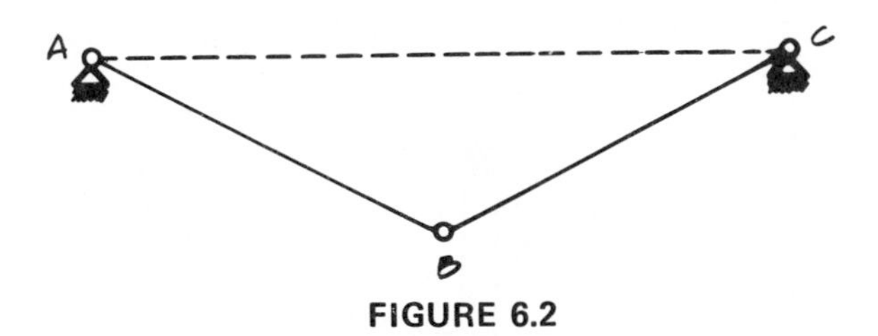

FIGURE 6.2

external constraints (the hinge) on the horizontal plane: The open door is externally unstable. Therefore, the moment produced by a pull on the door knob with the lever arm of the door width causes rigid rotation. If the door is locked, pounding on the door will produce bending deformation without rigid motion.

Two bars connected by an intermediate hinge and restrained at the ends by external hinges (Figure 6.2) are an example of a structure sufficiently constrained externally but internally unstable. The weight of the bars alone will produce rigid rotations of the bars.

A cable can be viewed as an infinite number of infinitesimal links connected by internal hinges. The terminal links are held in place by external hinges. A cable is therefore an internally unstable structure that can move rigidly to eliminate the lever arm of internal forces and, thus, the internal moments that it cannot carry. This feature of cable behavior will be reviewed as soon as external and internal cable forces have been completely defined.

In the first step of cable design (describing the structural scheme), the architect establishes the span according to measurements of the floor plan and the sag according to functional and aesthetic considerations. Such considerations include: head-room specification; optimization of the volume of internal space to be heated, cooled, and ventilated; lighting and acoustic specifications; and the shape of the roof line and interior spaces. The loads are then assessed. It is not necessary to specify constraints and the shape of the cable, since terminal sections as well as any section of the cable behave like hinges, and the shape is uniquely defined by span, sag, and load distribution.

In the second step, horizontal as well as vertical reactions must be evaluated. Experimental evidence for the existence of horizontal reactions is supplied, for example, by the behavior of a sling (Figure

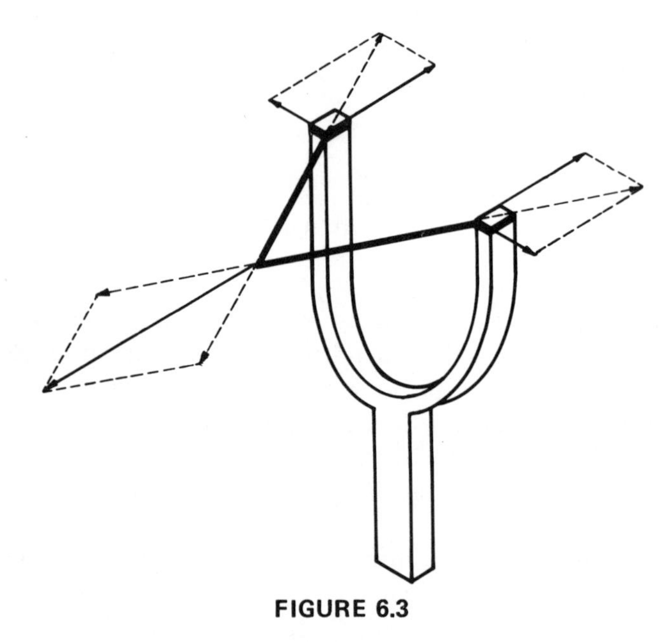

FIGURE 6.3

6.3). When the elastic band is loaded, the ends of the two arms of the sling move closer to each other. In the static condition, just before the slingshot is released, the action of the elastic band, which pulls the two arms inward, is balanced by the reactions of the arms, which keep the band stretched.

In the language of statics, we say that if vertical reactions alone supported a cable (Figure 6.4), bending moments would materialize in the form of products, such as $V_R\,d$, of reactions and their distances

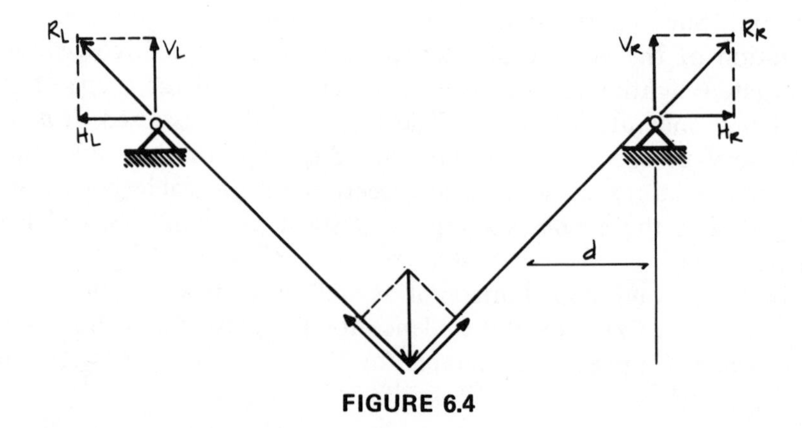

FIGURE 6.4

to various points of the cable. Since bending moments are zero everywhere on a cable, the reactions must be colinear with the cable's geometric axis. They therefore have both vertical and horizontal components.

The case of cables with supports at equal elevations and those with supports at different elevations are discussed separately.

6.1. CABLES WITH BOTH ANCHORS AT THE SAME ELEVATION

The cable of a suspension structure in Figure 6.5 is being designed to carry a set of vertical loads (for simplicity, concentrated at single points) over a span of 100 ft with a sag of 20 ft. The dead weight of the cable is considered negligible in comparison with the loads. Evaluating the reactions begins with solving an equation of moment equilibrium. Using, for example, the left end of the cable as pivot point, the equation is

$$\Sigma M = 0 = 100 V_r^b - 75(5) - 50(10) - 15(20),$$

from which

$$V_r^b = 11.75 \text{ k}.$$

Then, the equation of vertical equilibrium states

$$\Sigma F_y = 0 = V_l^b - 15 - 10 - 5 + 11.75,$$

and it yields

$$V_l^b = 18.25 \text{ k}.$$

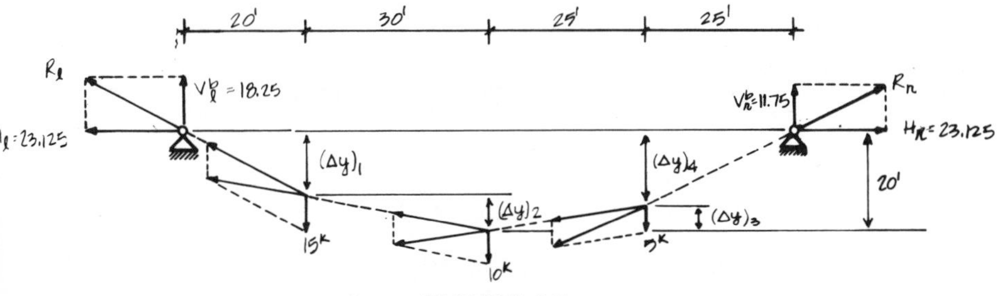

FIGURE 6.5

It should be noted that vertical reactions are calculated by the same approach used for beams and they have the same values as the vertical reactions of a beam with equal span, equal loads, and equal constraints. Therefore, we call V_l^b and V_r^b the vertical reactions of cables with both ends supported at the same elevation.

In the absence of horizontal loads, the equation of horizontal equilibrium is

$$\Sigma F_x = 0 = -H_l + H_r;$$

it simply yields

$$H_l = H_r.$$

Having used the three equations of equilibrium of external forces without finding the value of H_l or H_r, we must now resort to a different approach. Cables have the unique property of being bending free. We can, therefore, write in equation form that the bending moment is zero at any point of the cable for which we know the x and y coordinates. The specification of the sag gives the coordinate $\bar{y}$ of the cable's lowest point.

To find the $\bar{x}$-coordinate of this point, we recall that the moment diagram of a beam has the shape of the string polygon of the loading and reactive forces. The string polygon is similar to the shape of a string carrying the same loads, from which it takes its name. The cable, or string, therefore has the shape of the moment diagram of the *equivalent beam*, a beam having the same span, same load, and same constraints as the cable. Accordingly, the $\bar{x}$-coordinate of the cable's lowest point coincides with the coordinate of the beam section where the moment is largest and the shear is zero. The shear diagram (Figure 6.6) of the equivalent beam gives this coordinate, in our case, $\bar{x} = 50$ ft.

We can now set the bending moment $M_{\bar{x}}$, produced at the sag point by all forces on the left of the sag point, equal to zero

$$M_{\bar{x}} = 0 = -50(18.75) + 20H_l + 30(15),$$

from which

$$H_l = 23.125 \text{ k} = H_r.$$

All external forces are now available for Step 3, which, in the case

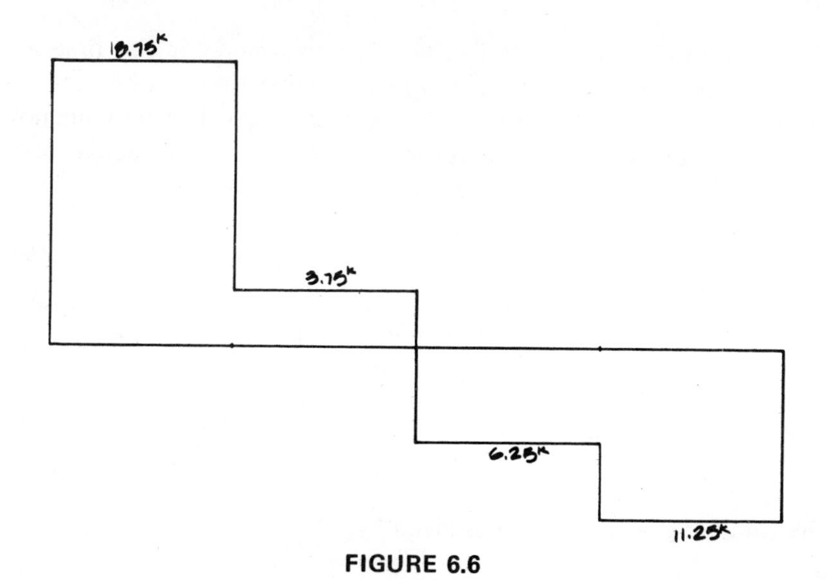

FIGURE 6.6

of cables, is the evaluation of the axial force in each segment and its slope.

The slope of the axial force is also the slope of the host cable segment, since they are colinear with one another.

In the example we are discussing, the combination of V_l and H_l for the left reaction gives

$$R_l = (V_l^2 + H_l^2)^{1/2} = (V_1^2 + H_1^2)^{1/2} = (18.25^2 + 23.125^2)^{1/2} = 29.4 \text{ k}.$$

The slope of R_l is

$$\tan \alpha_l = \tan \alpha_1 = \frac{V_l}{H_l} = \frac{V_1}{H_1} = \frac{18.25}{23.125} = 0.79 = \tan 38° \; 20',$$

which is also the slope of the first segment of the cable. With this slope, the cable's first segment descends a step $(\Delta y)_1$, given by

$$(\Delta y)_1 = (\Delta x)_1 \frac{V_1}{H_1} = 15.8 \text{ ft}.$$

The length of the first segment is (Figure 6.5)

$$L_1 = [(\Delta x)_1^2 + (\Delta y)_1^2]^{1/2} = 25.6 \text{ ft}.$$

The operations and results in Step 3, performed for the first segment of the cable, can be organized in a table that can be used for any cable carrying concentrated loads, including cables with anchors at different elevations. We therefore complete Step 3 in tabular form for the cable now being considered.

Segment i	V_i (k)	H_i (k)	$(V_i^2+H_i^2)^{1/2}$ (k)	V_i/H_i	$(\Delta x)_i$ (ft)	$(\Delta y)_i$ (ft)	L_i (ft)	$\Sigma(\Delta y)_i$ (ft)
1	+18.25	23.125	29.4	0.79	20	−15.8	25.6	−15.8
2	+3.25	23.125	23.4	0.141	30	−4.2	30.3	−20.0
3	−6.75	23.125	24.0	−0.292	25	+7.3	26.0	−12.7
4	−11.75	23.125	26.0	−0.508	25	+12.7	28.1	0.0
							$\Sigma L_i = 110$	

The total length of the cable is given by

$$\sum_{i=1}^{4} L_i = L_1 + L_2 + L_3 + L_4 = 110 \text{ ft.}$$

If a 110-ft string is anchored to two supports horizontally spaced 100 ft apart and if three loads weighing 15, 10, and 5 k are placed on the string at the respective horizontal coordinates of 20, 50, and 75 ft, the string will sag 20 ft at $\bar{x} = 50$ ft.

Figure 6.5 shows a graphic translation of the tabulated results. It also shows the flow of forces in the cable: The left reaction R_l is transferred by the first segment to the coordinate $x = 20$ ft, where it is combined with the 10-k load, and so on, to the opposite end of the cable and to the reaction R_r. This example should be used as a guide in solving various problems in this chapter.

At the conclusion of this chapter, a procedure is shown for obtaining internal forces and the cable shape more synthetically.

6.2. CABLES WITH ANCHORS AT DIFFERENT ELEVATIONS

The anchors at the two ends of a cable frequently have different elevations. In this case, horizontal reactions H_l and H_r, mutually equal in the absence of horizontal loads, are parallel rather than colinear.

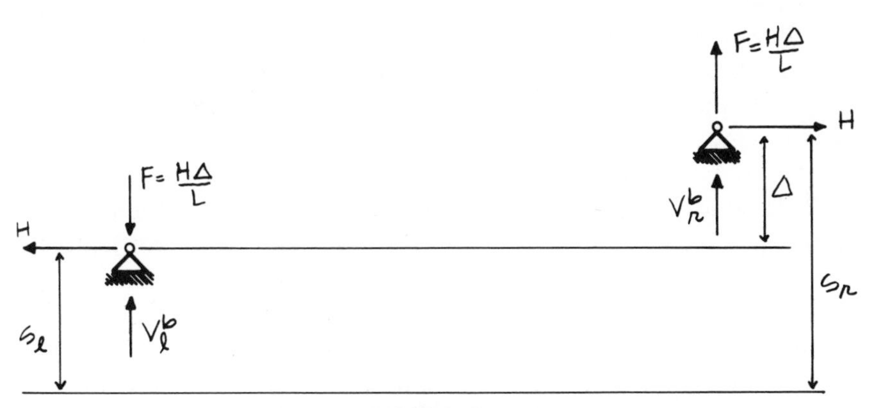

FIGURE 6.7

Calling s_l and s_r the sag of the cable measured from the elevation of the left and right supports, respectively, we say that the difference $\Delta = |s_r - s_l|$ is the arm of the couple of the horizontal reactions (Figure 6.7). Labeling these equal forces H for simplicity, the moment of their couple is $H\Delta$. The vertical reactions V_l and V_r must balance the moment $H\Delta$ as well as the vertical loads.

The share V_l^b of V_l and V_r^b of V_r required to balance the vertical loads are identical to the reactions of a cable with anchors at equal elevations. These reactions V_l^b and V_r^b are, in fact, calculated by equilibrium equations containing vertical forces and horizontal arms, neither of which is altered by a vertical relative displacement of the anchors. The share F (Figure 6.7) of V_l and V_r required to balance the moment $H\Delta$ must respect the condition

$$lF - H\Delta = 0,$$

from which

$$F = \frac{H\Delta}{l}.$$

The total vertical reactions are, therefore,

$$V_l = V_l^b - \frac{H\Delta}{l}$$

$$V_r = V_r^b + \frac{H\Delta}{l}.$$

The sign of $H\Delta/l$ is the same as that of V_r^b at the higher support; it is opposite the sign of V_l^b at the lower support, which we are assuming to be at the left end of the cable. At this end, therefore, $H\Delta/l$ is negative (downward).

When $H\Delta/l$ is less than V_l^b (Figure 6.8), the vertical component V_l of R_l is positive (upward). The diagram of the vertical components of the axial forces crosses its base line at a point (sag point) placed between the supports. The cable, in this case, could be said to be *relaxed*.

When $H\Delta/l$ is greater than V_l^b, V_l is negative (Figure 6.9). The diagram of the vertical components of the axial forces is all on one side of its base line, which means that the sag point can only be ideally placed externally to the lower support, where the extension of the

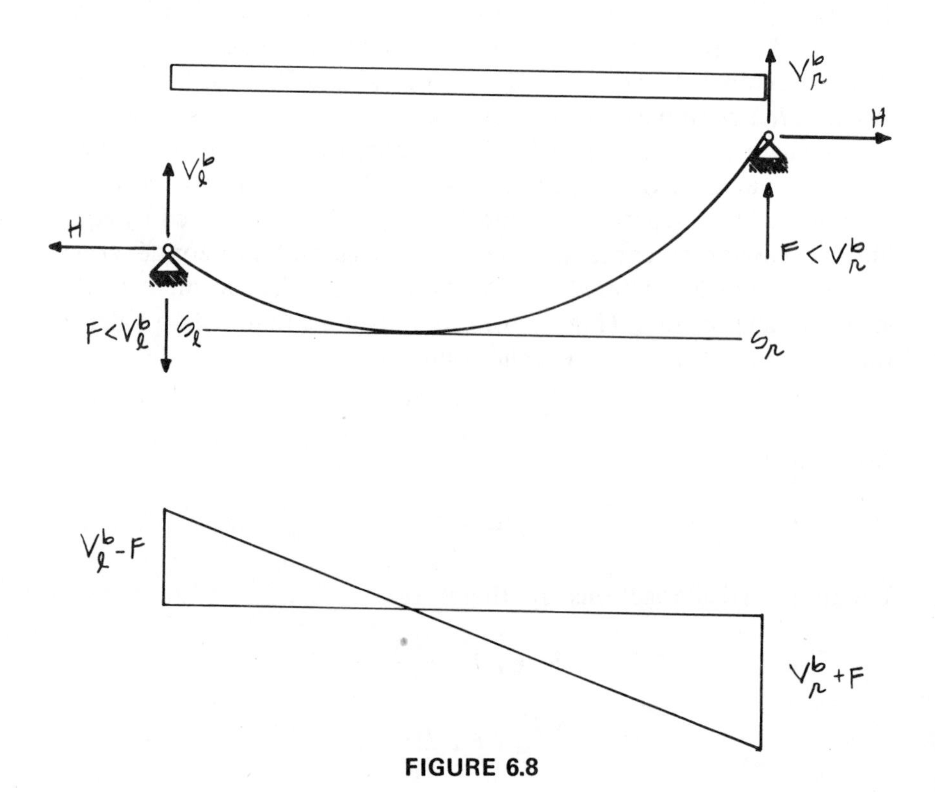

FIGURE 6.8

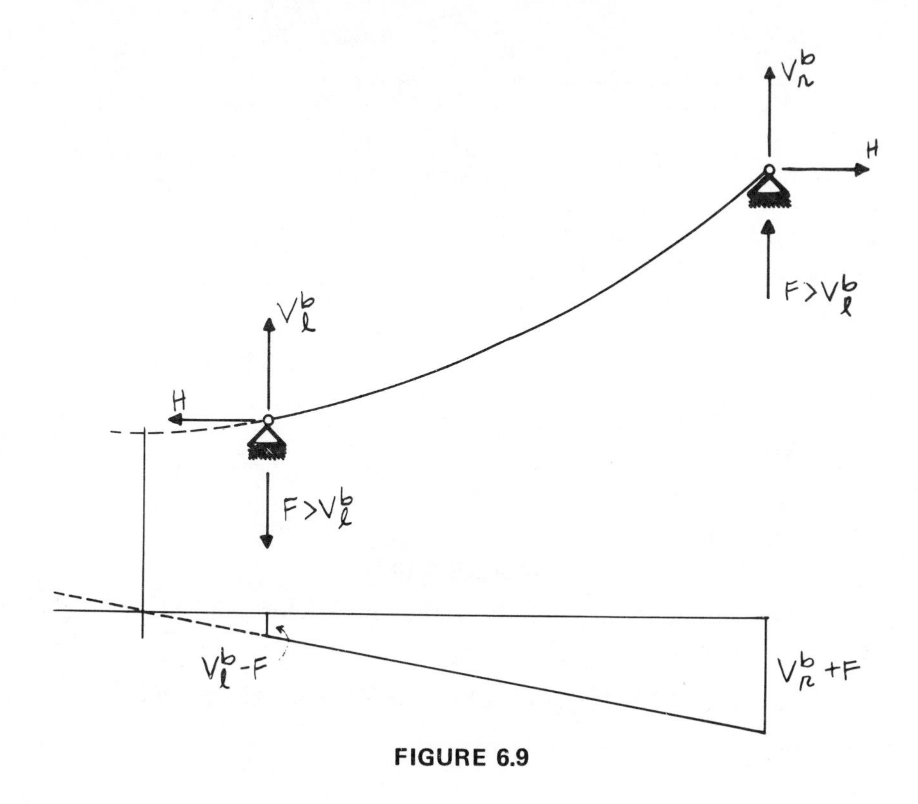

FIGURE 6.9

vertical force diagram crosses the base line. In this case, the cable could be said to be *taut*.

When $V_l^b = H\Delta/l$, $V_l = 0$, and the sag point coincides with the lower support (Figure 6.10). In this case, the cable is said to be *upper balanced* to indicate that all loads on the cable are balanced by the vertical reaction of the upper support.

It is the architect's responsibility to choose the values of the sag s_l and s_r that determine the occurrence of one of the preceding cases rather than others. We will investigate the case of the upper-balanced cables first because it is the easiest one.

Step 1. The loads shown in Figure 6.11 are distributed on a cable that spans 120 ft and has a sag $s_r = 12$ ft and sag $s_l = 0$ according to the architect's specifications.

Step 2. To obtain the specified geometric condition, the reaction V_l must be zero. Its fraction V_l^b is given by the equation that sets the

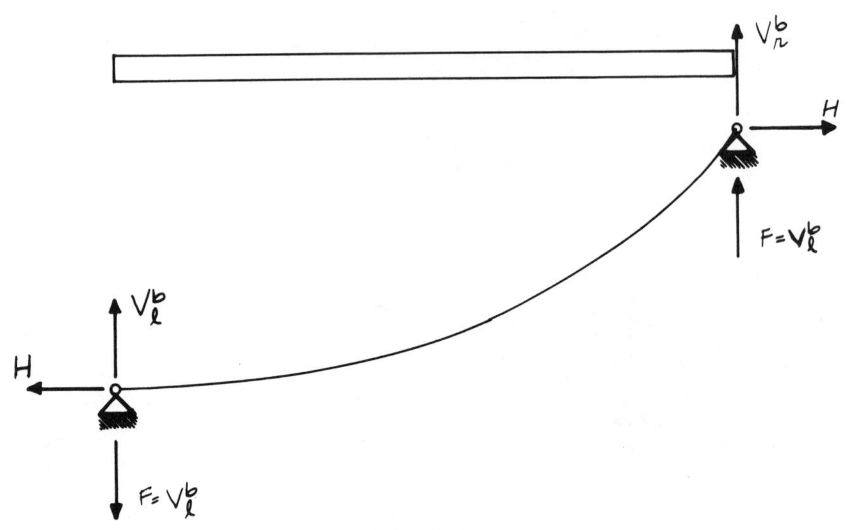

FIGURE 6.10

sum of all moments around the right end of the equivalent beam equal to zero

$$\Sigma M = 0 = -120V_l^b + 50(100) + 120(90) + 60(40) + 50(20),$$

from which

$$V_l^b = \frac{1}{120}(5000 + 10{,}800 + 2400 + 1000) = 160 \text{ k}.$$

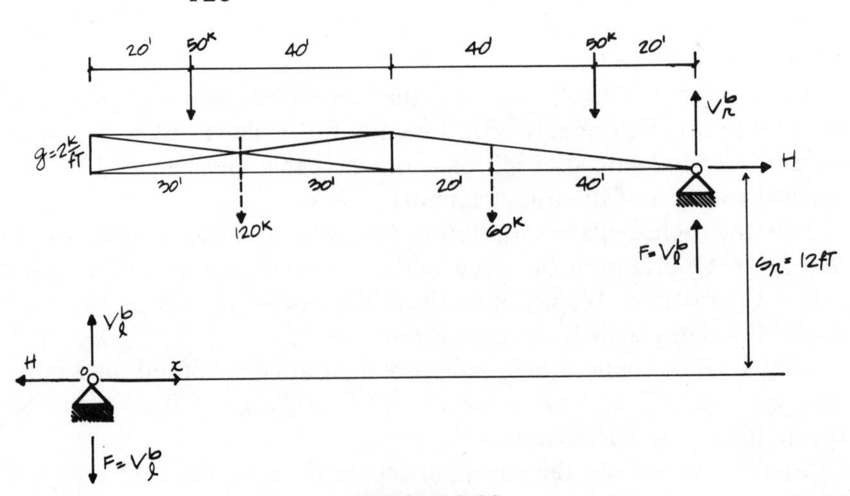

FIGURE 6.11

The horizontal reaction H is now obtained by setting V_l equal to zero

$$V_l = V_l^b - \frac{H\Delta}{l} = 160 - \frac{12H}{120} = 0,$$

from which

$$H = 1600 \text{ k}.$$

Step 3. Using the resultant 120 k of the uniform load and the resultant 60 k of the linear load diagram, we obtain approximate values for the internal forces and the cable geometries following the steps in the preceding table or Figure 6.5. The actual load diagrams must be used to obtain precise values for the axial forces and cable geometries.

For example, in the concentrated-load approximation, the axial force at $x = 15$ ft is $H = 1600$ k. This force is horizontal; therefore, the cable is horizontal at $x = 15$ ft. In reality, the axial force at $x = 15$ ft is obtained by combining the horizontal component $H = 1600$ k with the vertical component $V = 30$ k. Therefore,

$$P_{15} = (1600^2 + 30^2)^{1/2} = 1600.3 \text{ k}.$$

The slope of P_{15}, which is also the slope of the cable at $x = 15$ ft, is

$$\frac{V}{H} = \frac{30}{1600} = 0.019$$

rather than zero.

The equation that gives the vertical coordinate y *for any value of the horizontal coordinate* x *is obtained by setting the bending moment in the cable at the* x-*coordinate equal to zero.*

The equation has a different form in different regions of the load diagram.

For example, in the region of the load diagram limited by V_l and the 50-k load ($0 < x < 20$)

$$M(x) = 0 = -yH + gx\left(\frac{x}{2}\right),$$

from which

$$y = \frac{gx^2}{2H} = \frac{x^2}{1600}.$$

Therefore, at $x = 15$ ft, the vertical coordinate is

$$y = \frac{15^2}{1600} = 0.140625 \text{ ft or } 1.69 \text{ in.}$$

rather than zero, as in the concentrated load approximation. In the region defined by $20 < x < 60$,

$$M(x) = 0 = -yH + gx\left(\frac{x}{2}\right) + 50(x - 20),$$

from which

$$yH = \frac{gx^2}{2} + 50x - 1000$$

and

$$y = \frac{gx^2}{2H} + \frac{50x}{H} - \frac{1000}{H}.$$

We note that

The moment of H *around any point of the cable equals the moment produced by the loads.*

Reasoning graphically, we can say that

Due to a lack of bending strength, the sections of a cable move to points in space where the moment of H *equals the opposite moment of the loads, thus eliminating bending.*

In the region defined by $60 < x < 100$, the load is a linear function $g(x)$ of the x-coordinate

$$g(x) = g - \frac{g}{60}(x - 60) = 2g - \frac{gx}{60}.$$

Setting the bending moment equal to zero, we obtain

$$M(x) = 0 = -yH + 50(x - 20) + gx\left(\frac{x}{2}\right)$$

$$- \frac{g}{60}(x - 60)\frac{(x - 60)}{2}\frac{(x - 60)}{3},$$

expanding,

$$yH = \frac{gx^3}{360} + \frac{gx^2}{6} + 50x + 20gx - 600g - 1000.$$

Therefore,

$$y = \frac{x^3}{180H} + \frac{x^2}{3H} + \frac{90x}{H} - \frac{2200}{H}.$$

At $x = 60$ ft, the preceding equation yields

$$y = 3.5 \text{ ft}.$$

The coordinate $x = 60$ ft also belongs to the range of the uniform load. Therefore, the same value of y can be obtained from the quadratic equation.

The reader may want to set up the shape equation in the range of $100 < x < 120$ for additional practice. The equation will be cubic again, since the distributed load is still a linear function of x. It should be noted that the shape equation, obtained by setting the bending moment equal to zero, has in every load range the same degree as the moment curve for that range. This is an additional confirmation of the statement that the diagram of moments due to a given load distribution has the shape of a string carrying the same loads.

The procedure outlined for setting up the shape equation is general and can be used without modification for any cable carrying distributed loads, provided that the origin of the x- and y-coordinates is placed at the sag point.

In the case of polygonal shapes, the length of the cable is obtained by adding the tabulated lengths L_i of all sides of the cable polygon. In the case of distributed loads, the cable has the shape of curves whose equations are obtained by setting the bending in various regions of the load diagram equal to zero. The cable length can not be calculated, therefore, in finite terms but rather by integrating respective expressions of an element ds of cable segments, region by region.

In the following example, the length of the cable in Figure 6.11 is calculated in the region $0 < x < 20$. The procedure, indeed, is general and can be applied to other regions of the cable and to other cables. In the region considered, the equation of the cable curve is

$$y = \frac{gx^2}{2H} = \frac{x^2}{1600}.$$

The equation of the slope of the cable is

$$\frac{dy}{dx} = \frac{gx}{H} = \frac{x}{800}.$$

The expression of the element ds of this cable segment is, therefore,

$$ds = (dx^2 + dy^2)^{1/2} = dx\left[1 + \left(\frac{dy}{dx}\right)^2\right]^{1/2} = dx\left[1 + \left(\frac{gx}{H}\right)^2\right]^{1/2}.$$

The length of the cable in the region $0 < x < 20$ is

$$L_1 = \int_0^{20} ds = \int_0^{20} dx\left[1 + \left(\frac{gx}{H}\right)^2\right]^{1/2}$$

$$= \int_0^{20} dx\left[1 + \left(\frac{x}{800}\right)^2\right]^{1/2} = 20.03 \text{ ft.}$$

The same load distribution considered in the upper-balanced cable is now used to discuss the case of a *relaxed cable*.

Step 1. The values chosen for the left and right sags are (Figure 6.12)

$$s_l = 12 \text{ ft}, \qquad s_r = 24 \text{ ft}.$$

Step 2. Reactions V_l^b and V_r^b of the equivalent beam do not change because loads and span remain the same, as in the previous case. It is evident from the expressions for V_l and V_r

$$V_l = V_l^b - \frac{H\Delta}{l}, \qquad V_r = V_r^b + \frac{H\Delta}{l}$$

that the diagram of the vertical components of the axial forces can not be drawn before H is evaluated. The coordinate $\bar{x}$ of the sag point, which is obtained from that diagram, is, therefore, unknown, as is H.

To obtain values for $\bar{x}$ and H, we write two equations, one of which sets the vertical component $V(\bar{x})$ of the axial force equal to zero at $\bar{x}$, and the other sets the bending moment $M(\bar{x})$ equal to zero at $\bar{x}$

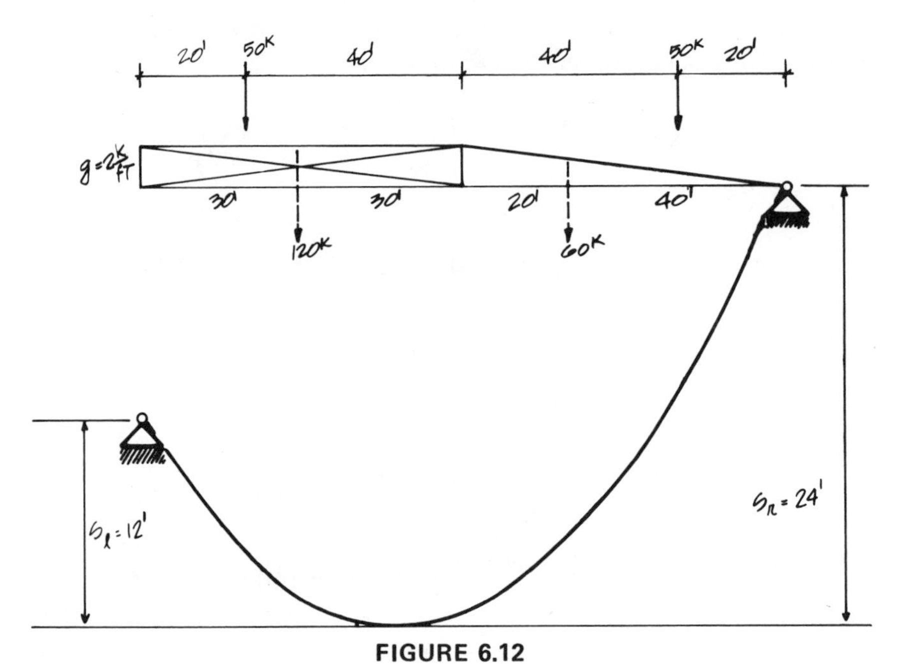

FIGURE 6.12

$$V(\bar{x}) = 0, \qquad M(\bar{x}) = 0.$$

It has been shown in the previous case that moment $M(\bar{x})$ and shear $V(\bar{x})$ have various expressions depending on the region of the load diagram in which $\bar{x}$ is located. Since the region in which $\bar{x}$ falls is unknown, it is evident that the problem must be solved by trials.

In the first trial, the range of x-coordinates in which $\bar{x}$ is expected to fall is assumed. The equations are written and solved. If $\bar{x}$ belongs to the assumed range, the results are correct, and additional trials are not necessary; otherwise, a new range must be assumed. It is seldom necessary to resort to a third trial, since the first trial gives an indication of the range to be chosen next. If in any trial the $\bar{x}$-coordinate is assumed to be that of a concentrated load, the equation $V(\bar{x}) = 0$ becomes redundant, since $\bar{x}$ has been fixed, and H is the only unknown. This is necessarily the case when the cable carries only concentrated loads.

We now proceed to evaluate the horizontal reaction H, and we assume in the first trial that

$$\bar{x} = 20 \text{ ft},$$

which is the coordinate of the 50-k load. Then,

$$M(\bar{x}) = 0 = s_l H - \bar{x} V_l^b + \bar{x} \frac{H\Delta}{l} + g\bar{x}\left(\frac{\bar{x}}{2}\right).$$

Numerically,

$$M(\bar{x}) = 0 = 12H - 20(160) + 20(12)\frac{H}{120} + \frac{2}{2}(20)^2,$$

from which

$$H = \frac{1}{14}(3200 - 400) = 200 \text{ k}$$

$$\frac{H\Delta}{l} = \frac{12(200)}{120} = 20 \text{ k}$$

$$V_l = V_l^b - \frac{H\Delta}{l} = 160 - 20 = 140 \text{ k}.$$

At $\bar{x}$, the vertical component of the axial force has a 50-k discontinuity due to the 50-k load. On the left side of the concentrated load,

$$V(\bar{x}) = V_l - g\bar{x} = 140 - 2(20) = 100 \text{ k}.$$

On the right side of the concentrated load,

$$V(\bar{x}) = V_l - g\bar{x} - 50 = 50 \text{ k}.$$

The diagram of $V(x)$ does not cross its base line at $\bar{x}$ as we had assumed. A new trial is, therefore, initiated, assuming that $\bar{x}$ will fall in the region $20 < x < 60$. Then,

$$V(\bar{x}) = 0 = V_l^b - \frac{H\Delta}{l} - g\bar{x} - 50,$$

$$M(\bar{x}) = 0 = s_l H - \bar{x} V_l^b + \frac{\bar{x} H\Delta}{l} + g\bar{x}\left(\frac{\bar{x}}{2}\right) + 50(\bar{x} - 20).$$

It should be noted that the moment equation is obtained by multiplying the forces in the shear equation with their respective arms and adding the moment of H with its correct sign.

Replacing letter symbols with their numerical values, we obtain

$$V(\bar{x}) = 0 = 160 - \frac{12H}{120} - 2\bar{x} - 50 = 110 - 2\bar{x} - \frac{H}{10},$$

$$M(\bar{x}) = 0 = 12H - 160\bar{x} + \frac{12H\bar{x}}{120} + \frac{2\bar{x}^2}{2} + 50\bar{x} - 1000,$$

$$M(\bar{x}) = 0 = \bar{x}^2 - 110\bar{x} + \frac{H\bar{x}}{10} + 12H - 1000.$$

From the shear equation, we obtain

$$H = 1100 - 20\bar{x}.$$

Substituting for H, the moment equation becomes

$$M(\bar{x}) = 0 = \bar{x}^2 - 110\bar{x} + \frac{\bar{x}}{10}(1100 - 20\bar{x})$$
$$+ 12(1100 - 20\bar{x}) - 1000.$$

Expanding and consolidating, we have the equation

$$\bar{x}^2 + 240\bar{x} - 12200 = 0.$$

The solution is

$$\bar{x} = 32.465 \text{ ft},$$

and it confirms that $\bar{x}$ falls in the range $20 < x < 60$. This value of $\bar{x}$ can, therefore, be used to obtain H.

$$H = 1100 - 20\bar{x} = 1100 - 20(43.1) = 238 \text{ k}.$$

Step 3 is performed as in the previous case and is not repeated here.

The case of a *taut cable* occurs when an architect specifies (Step 1) a cable shape with the sag point ideally located externally to the lower support.

For the purpose of evaluating the reactions, Step 2, we can turn this case into one of a relaxed cable by operating on the ideal extension of the real cable external to the lower support.

The ideal cable must continue with the same curve of the real cable at the lower support. The load diagram must, therefore, be extended externally to the lower support with the same shape it has at that end of the cable. The reactions of the ideal cable are equal and opposite to those of the real cable.

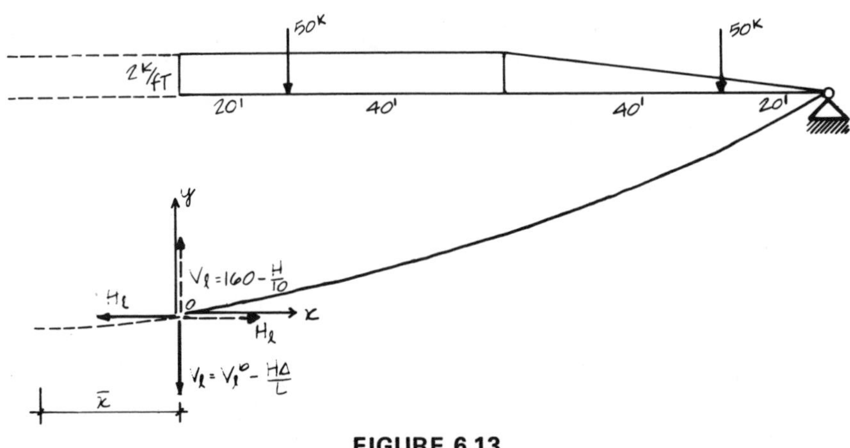

FIGURE 6.13

This approach is applied in the following example, where the span and load distribution in Figure 6.11 are used. The values specified for the sag are

$$s_l = 12 \text{ ft}, \qquad s_r = 24 \text{ ft}.$$

These are the elevations of the left and right anchors, measured from the horizontal tangent to the ideal extension of the cable curve external to the lower support. The real and ideal cables are shown in Figure 6.13, which also shows the extension of the load diagram and the reactions of the real and ideal cables at the lower anchor.

Working with the ideal relaxed cable, the reactions are evaluated by writing

$$V(\bar{x}) = 0 = V_l^b - \frac{H\Delta}{l} - g\bar{x},$$

$$M(\bar{x}) = 0 = \bar{x}V_l^b - \frac{\bar{x}H\Delta}{l} - s_l H - g\bar{x}\left(\frac{\bar{x}}{2}\right).$$

From the first equation,

$$H = (V_l^b - g\bar{x})\,\frac{l}{\Delta};$$

thus, the second equation becomes

$$M(\bar{x}) = 0 = \bar{x}V_l^b - \frac{\bar{x}\Delta}{l}(V_l^b - g\bar{x})\frac{l}{\Delta} - \frac{l}{\Delta}(V_l^b - g\bar{x})s_l - \frac{g\bar{x}^2}{2}.$$

Expanding and consolidating, we obtain

$$M(\bar{x}) = 0 = \frac{g\bar{x}^2}{2} + \frac{l}{\Delta}gs_l\bar{x} - \frac{lV_l^b s_l}{\Delta};$$

after multiplication by $2/g$ the equation, becomes

$$0 = \bar{x}^2 + \frac{2ls_l\bar{x}}{\Delta} - \frac{2lV_l^b s_l}{g\Delta}$$

The two solutions of this equation are

$$\bar{x} = \frac{ls_l}{\Delta}\left[-1 \pm \left(1 + \frac{2\Delta V_l^b}{gls_l}\right)^{1/2}\right]$$

We must choose the solution corresponding to a negative value of $\bar{x}$, since the sag point is to the left of the origin of the x-coordinates. Replacing the symbols with numerical values, we find that

$$\bar{x} = \frac{120}{12}(12)\left[-1 - \left(1 + \frac{2(12)160}{2(120)12}\right)^{1/2}\right] = -303.3 \text{ ft.}$$

The horizontal reaction is obtained by replacing this value of $\bar{x}$ in the expression for H

$$H = \frac{120}{12}\,[160 - 2(-303.3)] = 7666 \text{ k.}$$

The vertical reaction at the lower support is, therefore,

$$V_l = V_l^b - \frac{H\Delta}{l} = 160 - \frac{7666(12)}{120} = -606.6 \text{ k.}$$

The outward slope of the cable is

$$\frac{V_l}{H} = \frac{-606.6}{7666} = -0.079$$

If we imagine that the horizontal reaction of the lower anchor is supplied by a hydraulic jack and compare the results obtained for the taut cable with those for a relaxed cable with equal span, loads, and sags, we realize that the jack must pull with a 7666-k horizontal force to give the cable an outward slope of 7.9% at its lower support and a sag $s_l = 12$ ft external to the lower support. The same jack must produce only a 238-k horizontal pull to give the relaxed cable an inward slope

$$\frac{V_l}{H} = \frac{114.93}{450.7} = 25.5\%$$

at its lower support and a 12-ft sag s_l between supports.

The reason for the expressions taut cable and relaxed cable is now evident. The upper-balanced cable carrying the same loads on the same span has a horizontal reaction of intermediate magnitude

$$H = 1600 \text{ k}$$

and zero slope at the lower support.

Step 3 for taut cables follows the same procedure used for all other cables and is not repeated here.

6.3. GRAPHIC EVALUATION OF THE AXIAL FORCES AND GEOMETRIES OF A CABLE

A graphic approach to the operations in Step 3 is shown in Figure 6.5 for a cable with concentrated loads and anchors at the same elevation. The approach is, however, general and can be used for any support condition. If the loads are distributed rather than concentrated, they are first replaced by concentrated forces to obtain a polygonal cable shape. Using the sides of the polygon as tangents, we draw curves of appropriate degree where the cable carries distributed loads. A more synthetic approach uses a force polygon of the loads and reactions and the string polygon associated with those external forces.

Operating, for example, on the relaxed cable discussed in Section 6.1, we label the external forces as follows (Figure 6.14):

0–1 → the horizontal reaction H_l.

1–2 → the vertical reaction V_l.

2–3 → the 40-k load assumed concentrated at $x = 10$ ft, and so on, as shown in Figure 6.14.

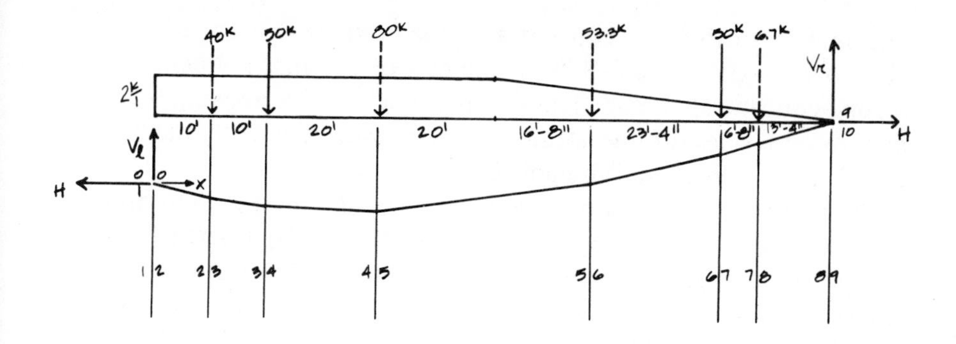

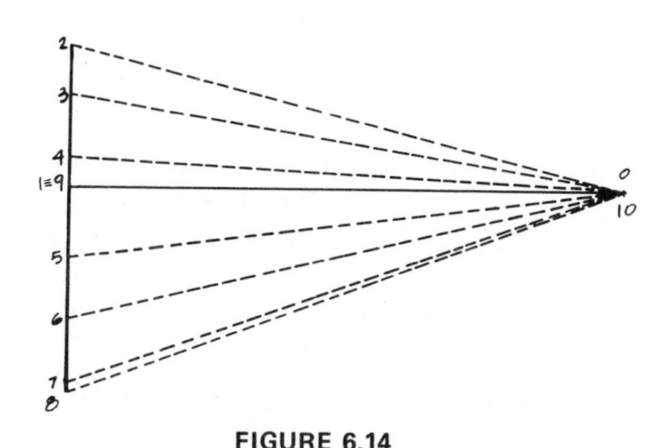

FIGURE 6.14

In the force polygon, these vectors are arranged in numerical order from 0–1 to 9–10. The axial forces in the cable are obtained by progressively summing the vectors of the force polygon.

For example, the axial force in the first segment of the cable is obtained by summing the vectors 0–1 and 1–2 according to the triangle rule. Their resultant summed to the next force, the 40-k load, gives the axial force in the next segment of the cable, and so on, to the right-end reaction R_r.

Since all the segments of the cable are colinear with the local axial force, we obtain the first segment of the cable by drawing a line parallel to the axial force 0–2, obtained on the force polygon, from the left anchor to the first concentrated load. We next draw a line parallel to the axial force 0–3 from the end of the first segment to the second concentrated load. These steps coincide with drawing the first and second sides of the string polygon associated with the forces 0–1, 1–2, . . . , 9–10 and the pole 0 = 10.

By completing this string polygon, we obtain the approximate shape of the cable. The precise shape is obtained by inserting curves of appropriate degree tangent to the sides of the string polygon where the cable carries distributed loads. Axial forces are measured with the scale S_F used to draw the force polygon. The sag at any point of the cable is measured with the scale used to draw the span l and the sags s_l and s_r.

PROBLEMS

6.1. The load diagram in Figure 6.15 is applied to a cable with anchors at the same elevation. The cable has a 280-ft span and a 30-ft sag. Replace the distributed loads with four concentrated forces. Find the vertical and horizontal reactions of the cable. Then, compile a table with the cable's internal forces and geometrical features.

Solution. The resultants of the distributed loads are:

A force F_1 = 160 k at x = 50 ft.
A force F_2 = 160 k at x = 110 ft.
A force F_3 = 60 k at x = 200 ft.
A force F_4 = 120 k at x = 250 ft.

The vertical reaction V_r^b of the right anchor is given by an equation of moment equilibrium with pivot at the left anchor

$$\Sigma M = 0 = -160(50) - 160(110) - 60(200) - 120(250) + 280V_r^b,$$

from which

$$V_r^b = \frac{1}{280}(8000 + 17{,}600 + 12{,}000 + 30{,}000) = 241.43 \text{ k}.$$

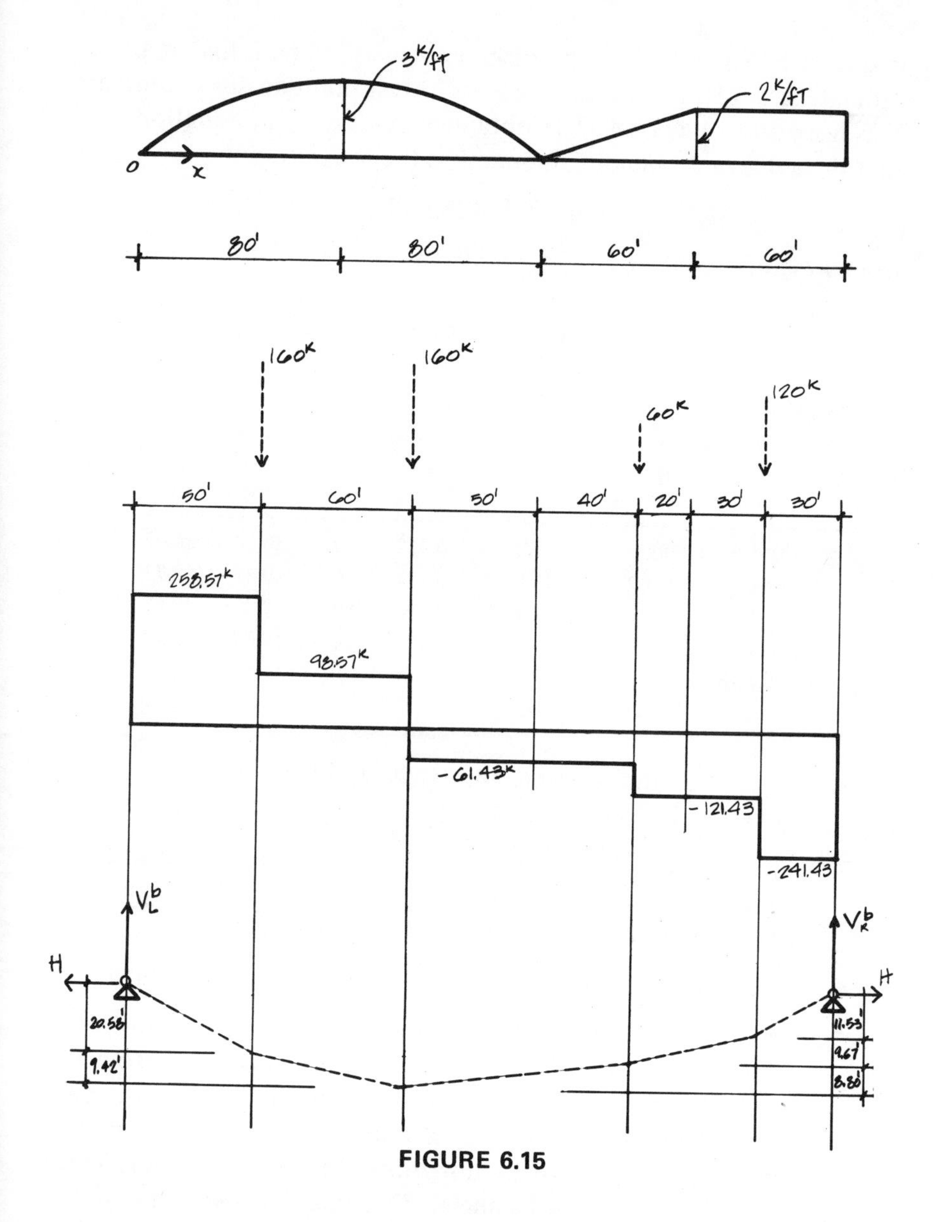

FIGURE 6.15

The equation of vertical equilibrium yields

$$V_l^b = 160 + 160 + 60 + 120 - 241.43 = 258.57 \text{ k}.$$

The diagram of the vertical components of the cable axial forces is shown in Figure 6.15. The diagram crosses its base line at the coordinate $x = 110$ ft; the cable has its lowest point at this coordinate. The horizontal reaction H is obtained by stating in equation form that the cable lacks bending moment at the lowest point.

$$M_{\bar{x}} = 0 = 30H - 110(258.57) + 160(60),$$

from which

$$H = \frac{1}{30}(28442.7 - 9600) = 628.09 \text{ k}.$$

The solution is completed in table form below.

Segment i	V_i (k)	H_i (k)	$(V_i^2 + H_i^2)^{1/2}$ (k)	V_i/H_i	$(\Delta x)_i$ (ft)	$(\Delta y)_i$ (ft)	L_i (ft)	$\Sigma(\Delta y)_i$ (ft)
1	258.57	628.09	679.2	0.412	50	−20.58	54.07	−20.58
2	98.57	628.09	635.8	0.157	60	−9.42	60.73	−30.00
3	−61.43	628.09	631.1	−0.098	90	+8.80	90.43	−21.20
4	−121.43	628.09	639.7	−0.193	50	+9.67	50.93	−11.53
5	−241.43	628.09	672.9	−0.384	30	+11.53	32.14	0
							288.30	

6.2. Following the graphic approach in Figure 6.5, obtain the same results as in Problem 6.1.

6.3. The load diagram in Fig. 6.16 is applied to a cable with anchors at the same elevation. The cable has a 300-ft span and a 36-ft sag. Perform the same tasks as in Problem 6.1.

6.4. Perform the tasks in Problem 6.2 for the cable in Problem 6.3.

6.5. The load diagram in Figure 6.15 is applied to a cable with anchors at different elevations. The sag measured from the left anchor is $s_l = 30$ ft. The sag measured from the right anchor is $s_r = 0$ ft. It is further specified that the tangent to the cable curve must be horizontal at the left anchor. Thus, the cable is upper balanced. Evaluate the reactions of the constraints.

Solution. By specification,

$$V_r = 0,$$

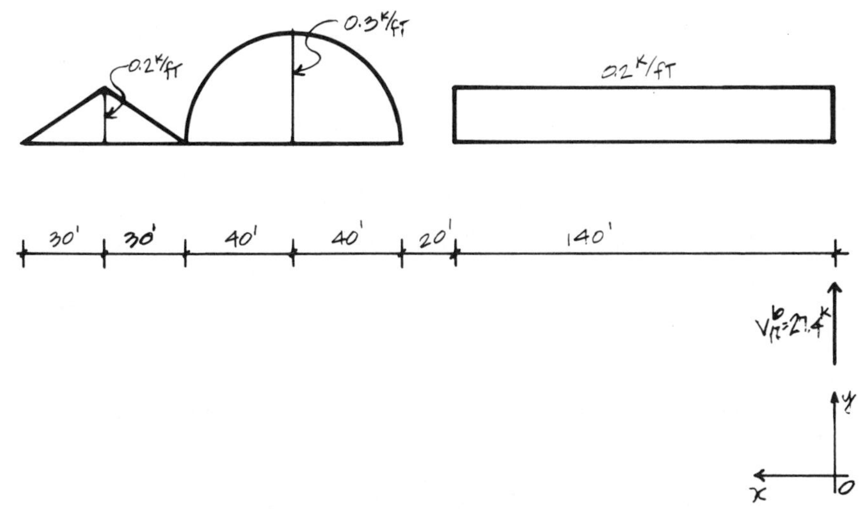

FIGURE 6.16

$$H = \frac{L}{\Delta} V_r^b.$$

From Problem 6.1, $V_r^b = 241.43$ k; thus,

$$H = \frac{280}{30}(241.43) = 2253.35 \text{ k},$$

$$V_l = V_l^b + V_r^b = 241.43 + 258.57 = 500 \text{ k}.$$

6.6. Compile the table of approximated axial forces and geometries for the cable in Problem 6.5, using the resultants of the distributed loads. Then, draw the correct cable shape with curves of appropriate degree tangent to the approximate string polygon.

6.7. Following the graphic approach in Figure 6.5, obtain the approximate polygonal shape for the cable in Problem 6.5.

6.8. Write the equation for the shape of the uniformly loaded part of the cable in Problem 6.5.

Solution. The origin of the x-coordinates is placed at the right anchor. At a point P of coordinates x and y $(0 < x < 60)$, the bending moment calculated from the right end is

$$M_p = 0 = yH - g\frac{x^2}{2}.$$

Thus,

$$y = \frac{gx^2}{2H} = \frac{2x^2}{2(2253.35)} = \frac{x^2}{2253.35} \text{(parabola)}.$$

6.9. The load diagram in Figure 6.16 is applied to an upper-balanced cable with $s_r = 0$ ft and $s_l = 36$ ft. Evaluate the reactions of the constraints. Using the resultants of the distributed loads, compile a table for the approximate axial forces and geometries, then draw the correct cable shape with curves of appropriate degree tangent to the approximate string polygon.

6.10. Following the graphic approach in Figure 6.5, obtain the approximate polygonal shape for the cable in Problem 6.9.

6.11. Write the equation for the shape of the cable in Problem 6.9 for $0 < x < 140$.

6.12. The load diagram in Figure 6.15 is placed on a relaxed cable with a sag $s_l = 30$ ft and a sag $s_r = 3$ ft. Evaluate the reactions of the constraints.

Solution. The origin of the x-coordinates is placed at the right anchor. The uniform load is labeled $g = 2$ k/ft. The beam reaction V_r^b is given in Problem 6.1; $V_r^b = 241.43$ k. It is not known in which load range the cable curve has a horizontal tangent. In the first trial, the assumption is, therefore, made that the tangent is horizontal at a coordinate $\bar{x}$ in the range $0 < \bar{x} < 60$. The vertical component of the axial force vanishes at $\bar{x}$; thus,

$$V_{\bar{x}} = 0 = V_r^b - \frac{H\Delta}{L} - g\bar{x}.$$

The bending moment vanishes at $\bar{x}$ (as everywhere else)

$$M_{\bar{x}} = 0 = \left(V_r^b - \frac{H\Delta}{L}\right)\bar{x} - Hs_r - g\bar{x}\left(\frac{\bar{x}}{2}\right).$$

From the first equation,

$$H = \frac{L}{\Delta}(V_r^b - g\bar{x}) = \frac{280}{27}(241.43 - 2\bar{x}) = 2503.72 - 20.74\bar{x}.$$

Replacing H in the second equation,

$$0 = V_r^b \bar{x} - \bar{x}\frac{\Delta}{L}(V_r^b - g\bar{x})\frac{L}{\Delta} - s_r \frac{L}{\Delta}(V_r^b - g\bar{x}) - \frac{g\bar{x}^2}{2};$$

expanding and consolidating,

$$0 = \frac{g\bar{x}^2}{2} + g s_r \frac{L}{\Delta}\bar{x} - V_r^b s_r \frac{L}{\Delta}.$$

Replacing symbols with their numerical values, this equation becomes

$$0 = \bar{x}^2 + 62.2\bar{x} - 751.2,$$

from which

$$\bar{x} = 60.98 \text{ ft}.$$

This value of $\bar{x}$ disproves the assumption that the cable has a horizontal tangent in the range $0 < \bar{x} < 60$. The value of $\bar{x}$ itself is, therefore, incorrect. From a practical viewpoint, however, the value $\bar{x}$ = 60.98 ft could be accepted. The precise value of $\bar{x}$ is found by solving the system of the shear and moment equations written in the next load range, where $\bar{x}$ most probably falls. These equations are

$$V_{\bar{x}} = 0 = V_r^b - \frac{H\Delta}{L} - g\bar{x} + \frac{g}{60}(\bar{x} - 60)(\bar{x} - 60)\frac{1}{2},$$

$$M_{\bar{x}} = 0 = \left(V_r^b - \frac{H\Delta}{L}\right)\bar{x} - Hs_r - g\bar{x}\left(\frac{\bar{x}}{2}\right) + \frac{g}{120}(\bar{x} - 60)^2\frac{(\bar{x} - 60)}{3}.$$

Extracting H from the shear equation and replacing symbols with numbers, we obtain

$$H = \frac{L}{\Delta}\left[V_r^b - g\bar{x} + \frac{g}{120}(\bar{x} - 60)^2\right] = \frac{280}{7}\left(301.43 - 4\bar{x} + \frac{\bar{x}^2}{60}\right),$$

from which

$$H = 0.17284\bar{x}^2 - 41.4815\bar{x} + 3125.94.$$

Replacing H in the moment equation, this becomes

$$M_{\bar{x}} = 0 = V_r^b \bar{x} - V_r^b \bar{x} + g\bar{x}^2 - \frac{g\bar{x}}{120}(\bar{x} - 60)^2$$

$$- \frac{Ls_r}{\Delta}\left[V_r^b - g\bar{x} + \frac{g}{120}(\bar{x} - 60)^2\right] - \frac{g\bar{x}^2}{2} + \frac{g}{360}(\bar{x} - 60)^3.$$

Expanding, consolidating, and dividing by $-2g$, the moment equation becomes

$$0 = \bar{x}^3 - \bar{x}^2\left(210 - 1.5\frac{L}{\Delta}s_r\right) - \bar{x}\left(-1800 + 360\frac{Ls}{\Delta}\right)$$

$$+ 180\frac{L}{\Delta}V_r^b\frac{s_r}{g} + 5400\frac{L}{\Delta}s_r + 168{,}000.$$

Replacing symbols with numbers, we obtain

$$0 = \bar{x}^3 - 163.33\bar{x}^2 - 9400\bar{x} + 952{,}004.$$

The solution is

$$\bar{x} = 60.88 \text{ ft},$$

which shows the close approximation of the first solution.

When $\bar{x}$ is replaced in the expression for H, we obtain

$$H = 0.17234(60.88)^2 - 41.4815(60.88) + 3125.94 = 1239.3 \text{ k}.$$

Then,

$$V_r = V_r^b - \frac{H\Delta}{L} = 241.43 - 1239.3\left(\frac{27}{280}\right) = 121.9 \text{ k},$$

$$V_l = V_l^b - \frac{H\Delta}{L} = 258.57 + 119.50 = 378.1 \text{ k}$$

6.13. For the cable in Problem 6.12, compile a numerical table of approximate axial forces and geometries by using the resultants of distributed loads.

6.14. Following the graphic approach in Figure 6.5 and the resultants of the distributed loads, obtain approximate axial forces and geometries for the cable in Problem 6.12.

6.15. The load diagram in Figure 6.15 is placed on a taut cable

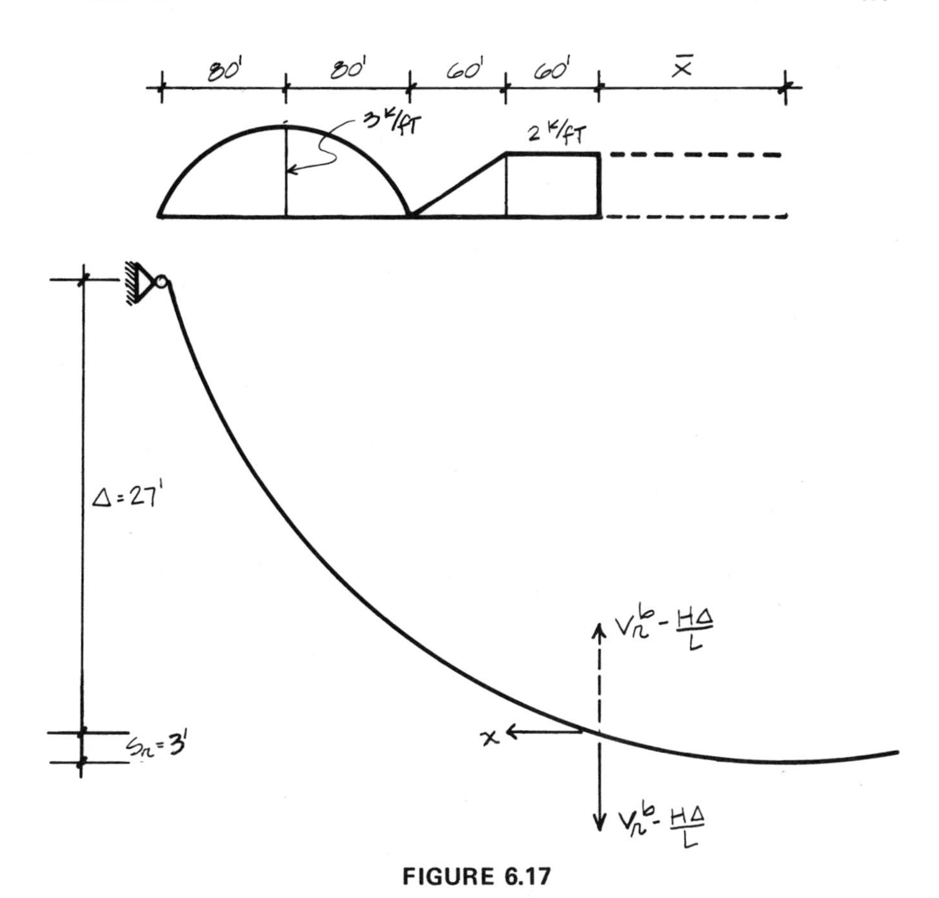

FIGURE 6.17

whose ideal sags are specified to be s_l = 30 ft and s_r = 3 ft. Evaluate the reactions of the constraints.

Solution. The load diagram is extended past the right end of the cable with the same distribution it has in the neighborhood of the right end (uniform). Figure 6.17 shows the ideal extension of the cable and its loads and the reactions of the ideal cable, which are opposite to the reactions of the real cable. The shear equation for the ideal cable is

$$V_{\bar{x}} = 0 = \left(V_r^b - \frac{H\Delta}{L}\right) - g\bar{x},$$

from which

$$H = \frac{L}{\Delta}(V_r^b - g\bar{x}) = \frac{280}{27}(241.43 - 2\bar{x}).$$

The moment equation for the ideal cable is

$$M_{\bar{x}} = 0 = -\left(V_r^b - \frac{H\Delta}{L}\right)\bar{x} + Hs_r + \frac{g\bar{x}^2}{2}.$$

Replacing H, we obtain

$$0 = -V_r^b\bar{x} + \frac{\Delta}{L}(V_r^b - g\bar{x})\frac{L}{\Delta}\bar{x} + s_r\frac{L}{\Delta}(V_r^b - g\bar{x}) + \frac{g\bar{x}^2}{2};$$

expanding, consolidating, and multiplying by $-2/g$, the equation becomes

$$0 = \bar{x}^2 + 2s_r\frac{L}{\Delta}\bar{x} - 2s_r\frac{L}{\Delta}\frac{V_r^b}{g}.$$

Replacing symbols with numbers, we obtain

$$0 = \bar{x}^2 + 62.2\bar{x} - 7511.2,$$

from which

$$\bar{x} = -31.1 \pm [(31.1)^2 + 7511.2]^{1/2} = -123.18 \text{ ft}.$$

The positive root gives an $\bar{x}$-coordinate between the anchors rather than outside of the right anchor, as prescribed and must be discarded. With this value of $\bar{x}$, we obtain:

$$H = \frac{280}{27}[241.43 - 2(-123.18)] = 5058.563 \text{ k},$$

$$V_r = V_r^b - \frac{H\Delta}{L} = 241.43 - 487.79 = -246.36 \text{ k (down)},$$

$$V_l = V_r^b + H\frac{\Delta}{L} = 258.57 + 487.79 = 746.56 \text{ k}.$$

6.16. For the cable in Problem 6.15, compile a numerical table of approximate axial forces and geometries by using the resultants of the distributed loads.

6.17. Using the graphic approach in Figure 6.5 and the resultants of the distributed loads, obtain approximate axial forces and geometries for the cable in Problem 6.15.

6.18. Using the graphic approach in Figure 6.14 and suitable scales s_F and s_D, draw the string polygon that approximates the shape of the cable in Problem 6.1. Then, insert the cable curves tangent to the string polygon. Indicate the degree of the cable curves in each load range.

6.19. Perform the tasks in Problem 6.18 for the cable in Problem 6.3.

6.20. Perform the tasks in Problem 6.18 for the cable in Problem 6.5.

6.21. Perform the tasks in Problem 6.18 for the cable in Problem 6.9.

6.22. Perform the tasks in Problem 6.18 for the cable in Problem 6.12.

6.23. Perform the tasks in Problem 6.18 for the cable in Problem 6.15.

Olympic Ice Center, Lake Placid, New York. Architects: Hellmuth, Obata, and Kassabaum, New York. Structural Engineers: Jack D. Gillum and Associates, St. Louis, Missouri. Photograph by George Cserna, New York.

SEVEN

TRUSSES

It has been said in discussing cable structures that optimal efficiency in using materials is realized by structures with geometries that allow loads to be transferred to supports by means of axial forces, which are the internal forces colinear with the axis of the structural elements. At the opposite end of the spectrum, structures that transfer loads by shear-bending mechanism are very inefficient because the material is underutilized almost everywhere on the typical section and along the span, as we easily recognize by observing the bending of a beam (Figure 7.1).

In discussing internal forces, we have compared a piece of a beam with an accordion and observed that bending produces maximum shrinkage and extension at the extreme fibers. Across the depth of the beam, compression gradually decreases, then reverses into tension, and increases as such to a maximum at the opposite extreme fibers. If the maximum moment produces tension and compression equal to their allowable values in the extreme fibers, all other fibers are stressed to a lesser degree and thus underutilized. On all sections where the moment is less than maximum, even extreme fiber stresses are less than their allowable values. In summary, only in the neighborhood of the top and bottom fibers at the section where bending is greatest does the material perform at capacity. Analogous reasoning applies to shear stresses.

Figure 4.2*b* shows that shear forces applied to an accordion, which simulates a piece of beam, increase or reduce the length of the diagonals of the side elevation. Fibers of material parallel to top and bottom fibers do not shrink or expand. In short, the usful zones of the prism of structural material are the top and bottom horizontal fibers, which carry the greatest bending stresses, and the two sets of diagonal fibers, which carry the greatest stresses produced by the

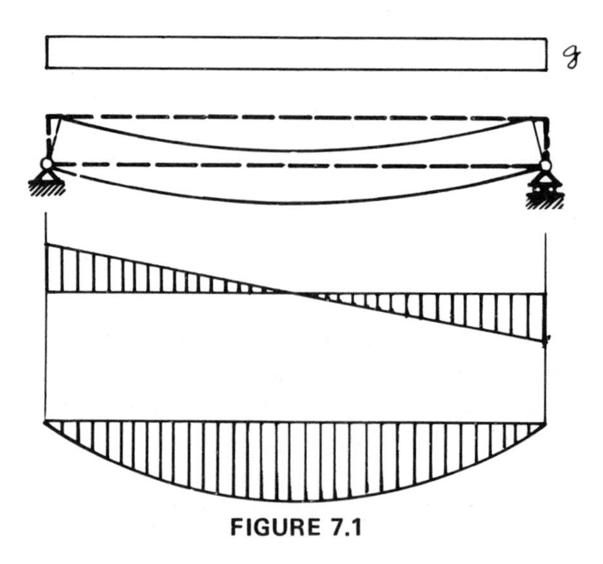

FIGURE 7.1

shear forces. If the idle material were removed, a more efficient structural element would result, as shown in Figure 7.2.

The same element could be fabricated by joining several bars in the same geometrical pattern. If the bars are connected by hinges rather than welds (Figure 7.3), each bar is moment-free at the ends and, therefore, all along its axis unless external loads are applied to the bar. Even in this case, bending moments are small because the span of an individual bar or link is small in comparison to the span of the entire structure.

We have, therefore, gradually accomplished the transition from the beam, which is an inefficient structural element due to its shear-bending mechanism, to the *truss*. In this new structure, tensile and compressive stresses due to bending and shear are all gathered in the

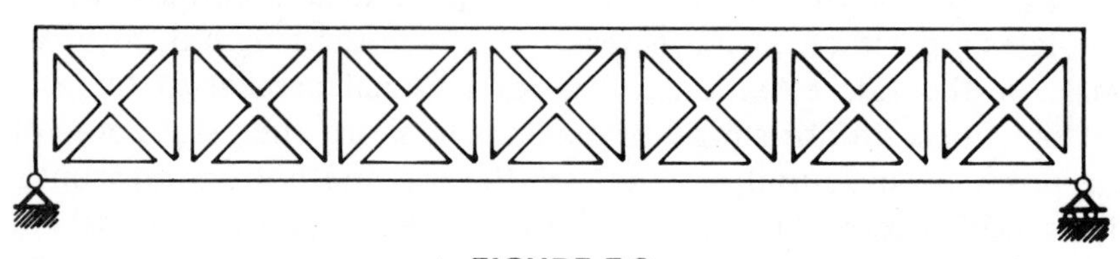

FIGURE 7.2

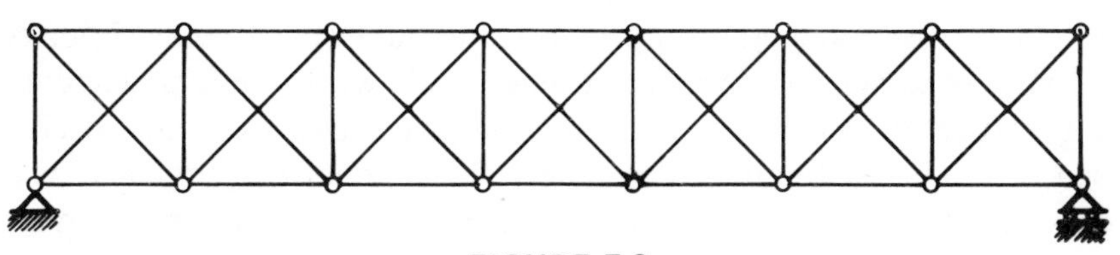

FIGURE 7.3

channels where they were flowing with their maximum intensity even before the idle channels were eliminated altogether. The selected channels are, of course, the bars of the truss; the idle channels are the understressed fibers of the beam.

If the joints connecting the bars are not welded and the loads are all concentrated at the joints rather than applied to the bars, bending and shear are completely eliminated in the structure. The only internal forces are axial tension or compression in the bars. The structure is not so efficient as a cable, since compression is always associated with the risk of sudden instability of the state of equilibrium, while tension, on the contrary, is always associated with stability. This is readily seen when a nail buckles under the compressive blows of a hammer. A nail, of course, would never buckle while being pulled by pliers.

Trusses, however, have the advantage of rigidity, which cables lack. Architects and engineers therefore resort to trusses when a large span or heavy load would require a hefty flexural member where large amounts of material would be underutilized and the flexibility of cables is not desirable.

With the appropriate geometries, trusses can perform the function of any element of structural framing: joists, girders, columns, and arches. Space trusses, extended out of one plane, can be found in the form of horizontal roofs and floors, folded roofs, barrel vaults, domes, and tubes carrying horizontal as well as vertical loads of tall buildings.

Most construction materials have been used to fabricate trusses. Wood was used almost exclusively for the bars of majestic roof trusses in ancient halls and cathedrals; wooden trusses are used today for buildings of modest scope. Modern structures most frequently use steel trusses ranging from prefabricated long-span open-web joists and girders to spectacular space-frames made with standard elements or specifically designed for exceptional buildings. Concrete is also used, but not so frequently as steel. The number of geometric patterns used for trusses is practically limitless. A well-designed truss, however, should have geometries such that the compressive axial forces develop in the bars best fit to carry them.

For example, steel trusses have relatively slender bars, because steel is a strong and expensive material used in elements with relatively small cross section, which are capable of carrying exceptionally large axial forces under tension. Under compression, however, the slenderness of steel bars makes them prone to buckling. Enlarging the cross section of the bars and reducing the unrestrained length of the bars are two ways of reducing slenderness, but the first option is more costly. In a steel truss, therefore, compressive forces should be channeled into the short bars and tensile forces into the long bars. For this reason, the relative geometries of the diagonal bars of the truss and external forces should resemble those of cables and their loads.

In Figure 7.4, tension bars are shown with solid lines and compression bars with dotted lines. It could be said that bars *AB* and *BD* form a cable carrying load V_l and are supported by reactive forces F_1 and F_2. Bars *BD*, *DF*, and *FH* form a cable carrying loads F_2 and F_4 and are supported by reactions V_l and V_r. Bars *CE* and *EG* form a cable carrying load F_3, and so on. A different arrangement of the

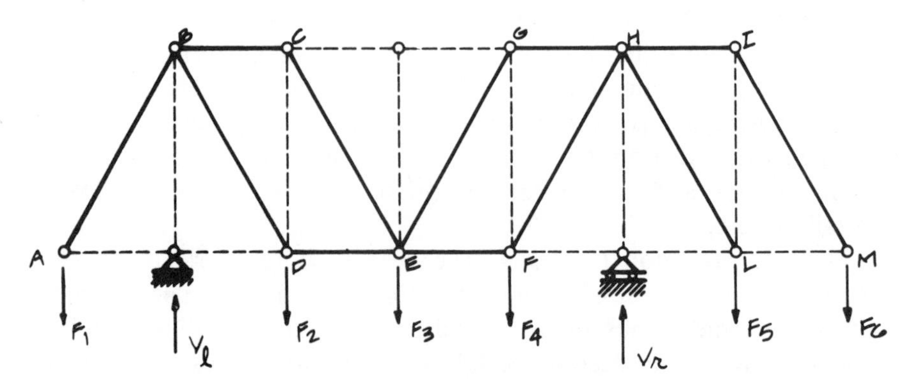

FIGURE 7.4

diagonal bars would inevitably produce compression in these long elements.

On the other hand, compression in the diagonal bars of a reinforced concrete truss is desirable in order to minimize the deflection of the truss. According to the definition of strain, which is

$$\epsilon = \frac{\Delta L}{L},$$

the elongation $\Delta L = \epsilon L$ of long diagonal bars is always greater than that of vertical bars. In the case of reinforced concrete trusses, the strain of the tension bars is really the strain of their reinforcing steel rods, since concrete has negligible tensile strength, and the steel must carry the tensile stresses.

Under working conditions, the strain of steel is greater than that of concrete. Using the notations

ϵ_s = strain in steel,
ϵ_c = strain in concrete,
P = axial force,
A = cross-sectional area,
$\frac{P}{A}$ = stress,
E = Young's modulus,

the strains in steel and concrete can, in fact, have the values

$$\epsilon_s = \frac{P}{A_s E_s} = \frac{30}{30{,}000} = \frac{1}{1000},$$

$$\epsilon_c = \frac{P}{A_c E_c} = \frac{1.5}{3000} = \frac{0.5}{1000} = \frac{\epsilon_s}{2}.$$

Then, tension in the reinforcement of the diagonals would extend them considerably because of the great original length and the great strain. As a result, deflections of the truss would be large.

To produce compression in the diagonals, the respective geometries of these bars and the external forces must resemble those of arches and their loads, as shown in Figure 7.5, where compression bars are drawn with solid lines. Bars AB and BC, for example, resemble an

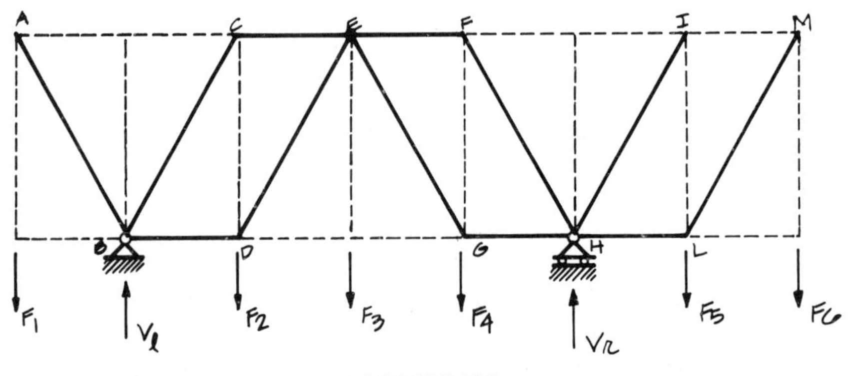

FIGURE 7.5

arch loaded by a force V_l and supported by reactive forces F_1 and F_2. Bars BC, CF, and FH resemble an arch carrying loads F_2 and F_4 and supported by reactions V_l and V_r, and so on. A considerable distortion of the truss, which occurs when the short bars become shorter and the long bars become much longer, is thus avoided.

Regardless of the structural material, deflections of trusses tend to be large when the diagonal bars form small angles with the top and bottom lines of the truss, called the top chord and the bottom chord, respectively. In the following example, the simplest of the trusses, a sign hanger formed by one diagonal and one horizontal bar (Figure 7.6*a*) substantiates our statement.

The gravity load G is decomposed according to the parallelogram rule into a compressive axial force C in the horizontal bar and a tensile force T in the diagonal bar. Therefore, the two bars shrink and expand, respectively. They then rotate with the angles α and β and are joined again at their hinged connection. The joint, as a result, has a vertical displacement whose fraction e is due to the extension of the diagonal bar and the fraction r to its rigid motion. A large value for the angle γ reduces the magnitude of components T and C of G (Figure 7.7*a*), thus the extension of the diagonal bar and its vertical component e (Figure 7.6*b*). A large value for the angle γ also reduces the vertical component r of the rigid displacement of the joint due to the rotation β. Indeed, with $\gamma = 90°$, the rotation β would not be required, and, in any case, it would not produce a downward displacement (Figure 7.7*b*). Of course, the diagonal bar would, in

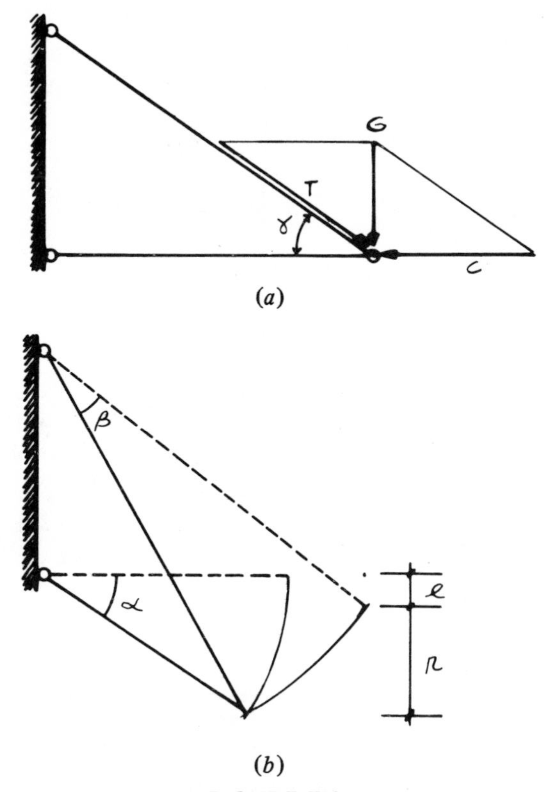

(*a*)

(*b*)

FIGURE 7.6

this case, become a vertical bar colinear with G, a structure as efficient as a cable. A good rule of thumb in designing horizontal trusses is to keep the angle γ in the 45–60° range.

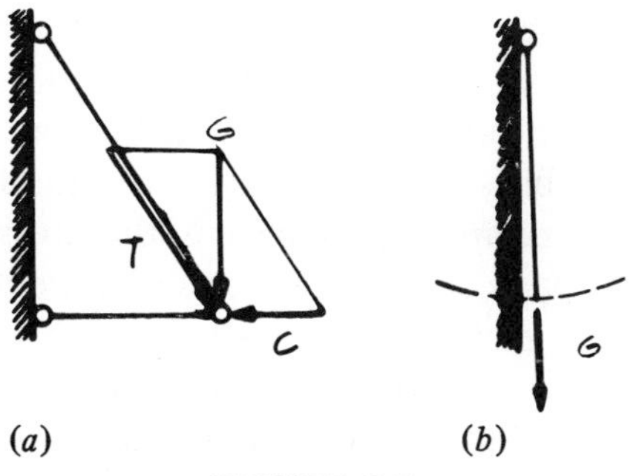

(*a*) (*b*)

FIGURE 7.7

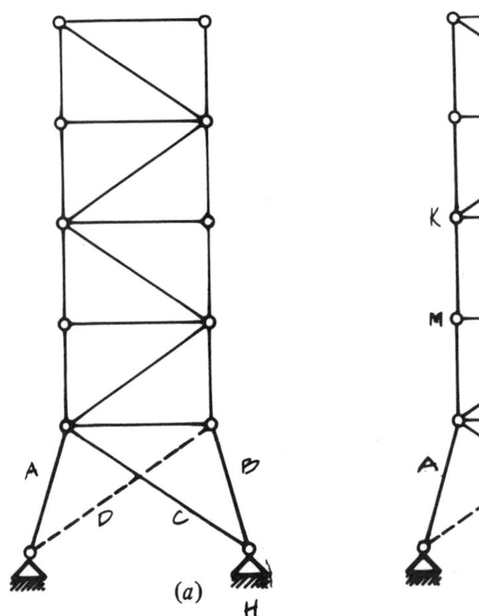

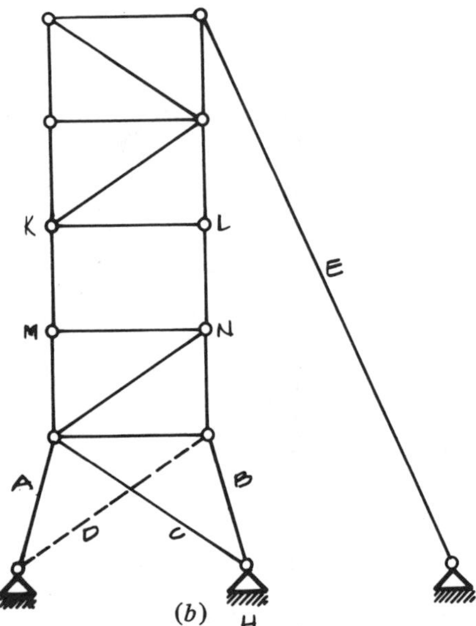

FIGURE 7.8

The constraints of a truss must be specified at Step 1, along with the geometries and loads. A truss is statically determinate if the number of constraints equals the number of potential movements (degrees of freedom) of its rigid parts. For example, the rigid truss in Figure 7.8*a* has three degrees of freedom on the *xy* plane and three constraining links *A*, *B*, and *C*. The constraining action of links *B* and *C* can also be identified with that of the external hinge *H*, whose reaction is the resultant of the reactions of *B* and *C*. If an additional link *D*, shown in Figure 7.8*a* by a dotted line, also constrains the truss, the reactions cannot be evaluated by equilibrium conditions. The truss has become statically indeterminate because the number of links exceeds the number of plane equilibrium equations.

As an additional example, we consider the truss in Figure 7.8*b*, which has two rigid parts connected by a slide, the deformable panel *KLMN*. The two rigid parts have six degrees of freedom. The sliding connection provides two constraints. The four external links *A*, *B*, *C*, and *E* provide the additional constraints required for stability. It is necessary to recognize that each rigid part of a truss must be sufficiently constrained, or the truss will be prone to collapse. If, for example, the dotted link D is used instead of link E, the rigid part above the collapsible panel $KLMN$ is not sufficiently constrained, and the lower rigid part is overconstrained.

Whenever possible, all panels of the truss should be made rigid by using a diagonal bar. Transferring loads by axial force mechanisms and, therefore, the efficiency of the truss are, in fact, conditioned by the geometric pattern suggested earlier in this chapter in the discussion of the transition from beam to truss. If two diagonal bars rather than one are used in a truss panel, axial forces cannot be obtained by conditions of equilibrium because the panel is overconstrained or redundant. In summary,

Plane trusses are statically determinate when they are made of rigid parts with one diagonal bar in each panel and the n *rigid parts are internally and externally connected by a number of constraints equal to the* 3n *degrees of freedom.*

The second part of this statement applies, of course, to any plane structure.

The operations in Step 2 are performed for a truss according to the general method for evaluating reactive forces, which involves setting up and solving equations of equilibrium of the truss after replacing the constraints with their unknown reactions. Two different methods are available for calculating axial forces in the bars of trusses (Step 3).

7.1. AXIAL FORCES IN TRUSSES BY THE RITTER SECTION METHOD

The method due to Ritter requires cutting the truss with a section across three nonconcurrent bars. A free body diagram is drawn for either one of the two parts divided by the Ritter section. In the diagram, the external forces (loads and reactions of the constraints) are known from Steps 1 and 2; the axial forces in the three bars split by the Ritter section are unknown, and they are evaluated by the three equations of plane equilibrium.

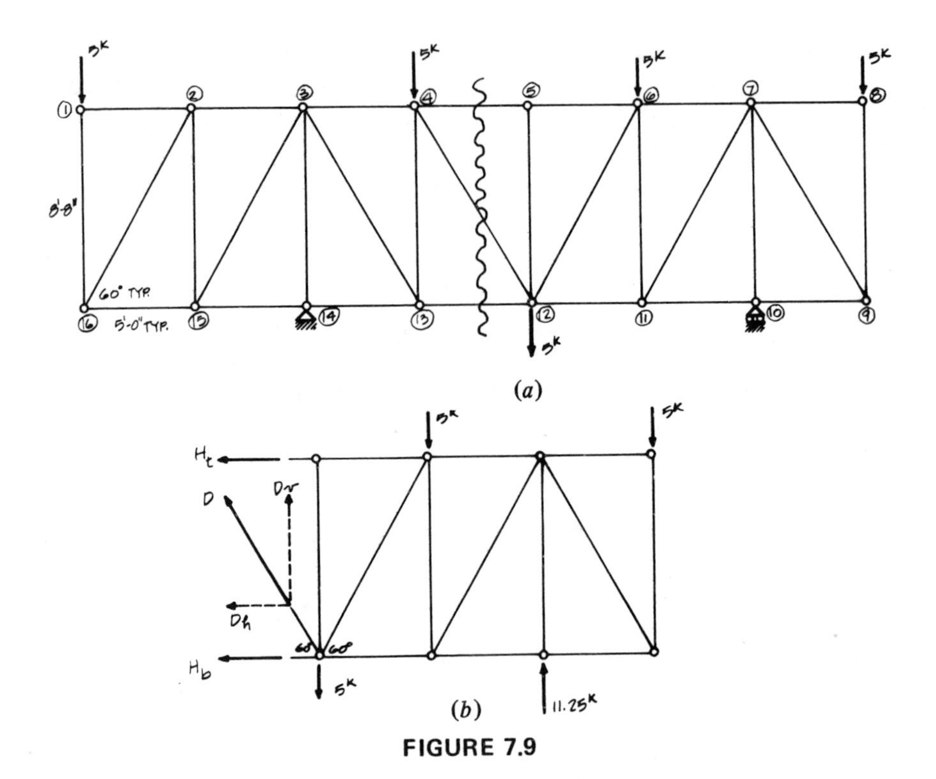

FIGURE 7.9

The following example illustrates the application of the Ritter method. The truss specified in Step 1 has the geometries, constraints, and loads shown in Figure 7.9*a*. The external force V_r is obtained by writing an equation of moment equilibrium around the hinge for the entire truss (Step 2).

$$\Sigma M = 0 = 5(10) - 5(5) - 5(10) - 5(15) + 20V_r - 5(25),$$

from which

$$V_r = \frac{1}{20}(25 + 75 + 125) = 11.25 \text{ k}.$$

The equation of vertical equilibrium yields the reaction V_l of the hinge

$$V_l = 13.75 \text{ k}.$$

A Ritter section S is used to evaluate axial forces in the three bars of one of the panels (Step 3). The free body diagram of the part to the right of S is shown in Figure 7.9*b*. The equation of vertical equilibrium states

$$\Sigma Fy = 0 = D_v - 5 - 5 + 11.25 - 5,$$

from which

$$D_v = 3.75 \text{ k}.$$

Then,

$$D_h = D_v \tan 30° = 2.17 \text{ k}.$$

The equation of moment equilibrium about the midspan joint of the bottom chord states

$$\Sigma M = 0 = 8.66H_t - 5(5) + 10(11.25) - 5(15).$$

The axial forces D and H_b as well as the 5-k midspan load, do not have lever arms around the chosen pivotal point. Solving the moment equation,

$$H_t = \frac{1}{8.66}(25 - 112.5 + 75) = -1.44 \text{ k}.$$

The minus sign in the result indicates that H_t was incorrectly assumed to be a tensile force.

After changing H_t from tensile to compressive force, the equation of horizontal equilibrium states

$$\Sigma Fx = 0 = 1.44 - 2.17 - H_b,$$

from which

$$H_b = -0.73 \text{ k}.$$

Again, the minus sign in the result shows us that H_b is not tensile, as assumed, but a compressive force instead. The same results can be found by operating on the part of the truss to the left of section S.

7.2. AXIAL FORCES IN TRUSSES BY THE JOINT METHOD

7.2.1. Numerical Evaluation

The other method used to evaluate axial forces in the bars of trusses is known as the method of equilibrium of the joints or, simply, the joint method. Each joint of a truss is subject to a system of concurrent forces, some of which are external forces like the loads specified in Step 1 and the reactions calculated in Step 2; and some are internal axial forces of the bars connected by that joint. The system is equilibrated, or else the joint would be set in motion. It is therefore correct to state that the sum of all forces is zero in both the x and y directions. The equation of moment equilibrium is useless because all the forces of the system converge to the joint; thus, they do not have lever arms around the joint.

The joint method uses the two equations of equilibrium

$$\Sigma Fx = 0,$$

$$\Sigma Fy = 0$$

to calculate two unknown bar forces at each joint of a truss. The other forces converging to the joint must be known loads, known reactions, or known bar forces evaluated by the equilibrium of neighboring joints.

In the joint method, the joints of the truss are numbered in any suitable order. The typical symbol used for the axial force in the bar that spans, for example, between joints 5 and 6 is $A_{5,6}$.

The axial force in a bar is compressive when the bar pushes on its terminal joints and it is, in return, pushed by the joints with equal and opposite forces, as is necessary for equilibrium (Figure 7.10a). The axial force in a bar is tensile when the bar pulls on its terminal joints and is, in return, pulled by them (Figure 7.10b).

Application of the joint method must start from a joint where only two axial forces are unknown; otherwise, two equations of force equilibrium are insufficient for evaluating all the unknown bar forces. For the same reason, application of the joint method must proceed to a joint in the neighborhood of the previous joint where

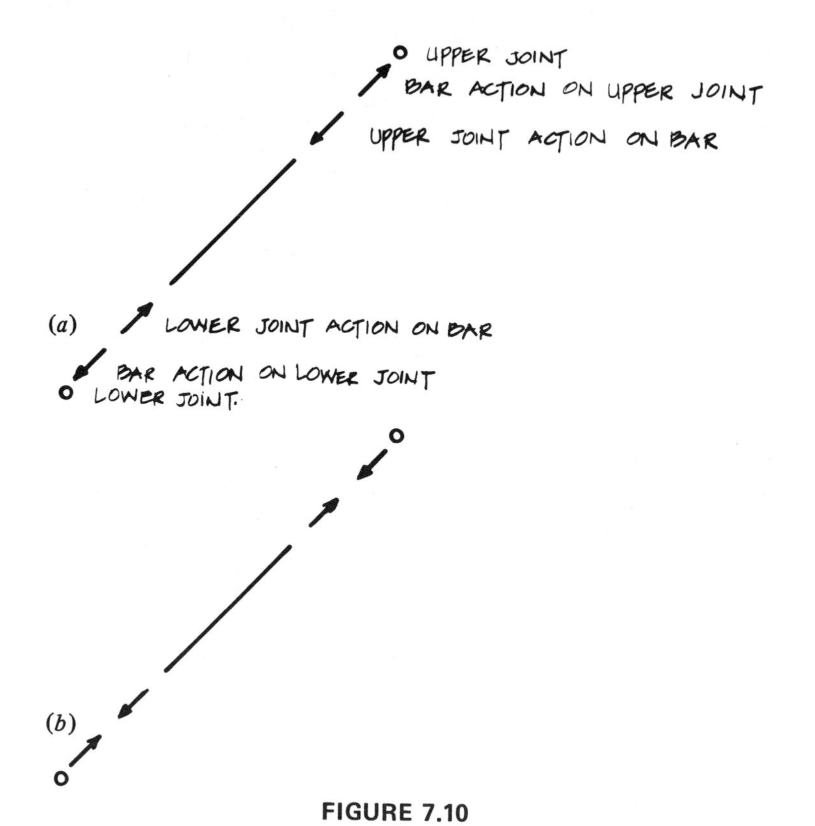

FIGURE 7.10

unknown axial forces have been reduced to only two in number. The truss in Figure 7.9*a* is used to demonstrate in detail how the joint method is applied.

Joints 1 and 8 are both suitable for starting the method. Starting from joint 1, we draw the free body diagram of the joint (Figure 7.11*a*), and we state

$$\Sigma Fx = 0 = A_{1,2},$$

$$\Sigma Fy = 0 = -5 + A_{1,16},$$

from which

$$A_{1,16} = +5 \text{ k}.$$

The positive sign in the results proves that $A_{1,16}$ pushes on its upper joint 1, as assumed in the free body diagram. $A_{1,16}$ is, therefore, a

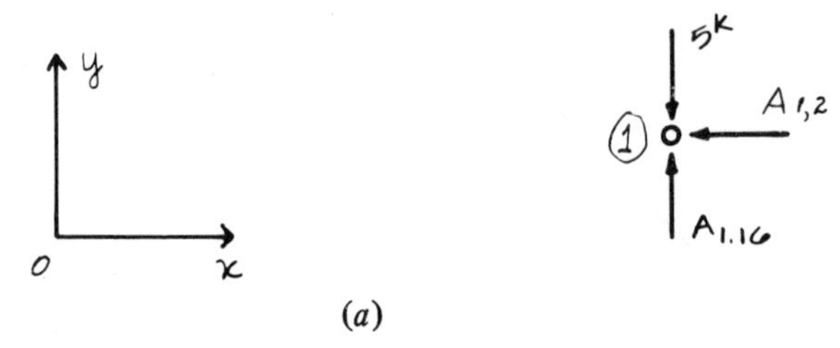

(a)

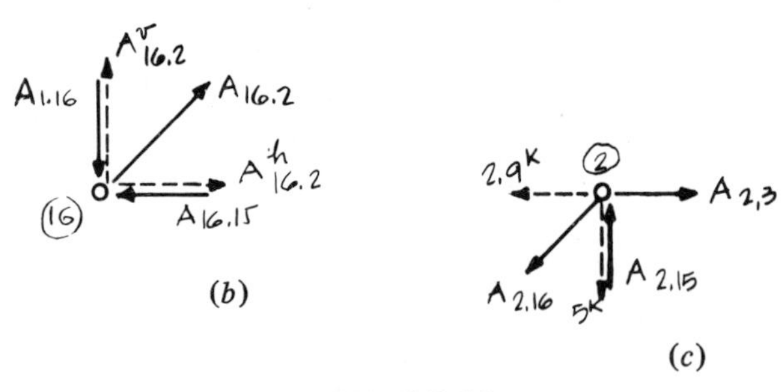

(b)

(c)

FIGURE 7.11

compressive force. The number of unknown bar forces at neighboring joint 16 is reduced to only two.

The free body diagram of joint 16 (Figure 7.11*b*) shows $A_{1,16}$ with a downward sign, since bar 1, 16 is in compression and pushes up on joint 1 and down on joint 16. The equations of equilibrium for joint 16 state

$$\Sigma Fx = 0 = A^{h}_{16,2} - A_{16,15},$$

$$\Sigma Fy = 0 = -5 + A^{v}_{16,2},$$

from which

$$A^{v}_{16,2} = 5 \text{ k}.$$

Then,

$$A^{h}_{16,2} = A^{v}_{16,2} \tan 30^{\circ} = 2.9 \text{ k}.$$

Substituting in the equation of horizontal equilibrium, we obtain

$$A_{16,15} = 2.9 \text{ k}.$$

The unknown bar forces at neighboring joint 2 have been reduced to only two. It is therefore possible to evaluate them by the two equations of horizontal and vertical force equilibrium.

The free body diagram of joint 2 is shown in Figure 7.11*c*. The equations of equilibrium are

$$\Sigma Fx = 0 = -2.9 + A_{2,3},$$

$$\Sigma Fy = 0 = -5 + A_{2,15},$$

from which

$$A_{2,3} = 2.9 \text{ k},$$

$$A_{2,15} = 5 \text{ k}.$$

According to the rule for identifying axial forces, bar 2,3 is in tension because it pulls on the joints, as the free body diagram assumes and the solution of the equation confirms. Bar 2,15 is in compression.

As the statement and solution of equilibrium conditions at each joint become repetitious, it is convenient to organize all calculations and results in a table. We therefore continue our investigation of axial forces in the bars of the truss in Figure 7.9*a* in Table 7.1. The 13 columns of the table contain the following items:

1. Free joint diagrams.
2. External forces: horizontal components.
3. External forces: vertical components.
4. Bars converging to the joint.
5. Known bar forces: horizontal components.
6. Known bar forces: vertical components. (The signs of the forces in columns 2, 3, 5, and 6 are positive if the forces are oriented like the positive x- and y-semiaxes.)
7. Unknown bar forces: horizontal components.
8. Unknown bar forces: vertical components.
9. First equilibrium equation. If only one unknown horizontal bar force (or component) exists and two vertical unknown bar forces (or components) exist, the first equation is $Fx = 0$ or $(7) = (-2) - (5)$. If the opposite case occurs, the first equation is $Fy = 0$ or $(8) =$

TABLE 7.1

1	2	3	4	5	6
	External Forces			Known Axial Forces	
Free Joint Diagram	Horizontal Component	Vertical Component	Bars	Horizontal Component	Vertical Component
5k, (1), A1.16	0	–5	1,2		
			1,16		
5k, A16.2, (16), A16.15			16,1		–5
			16,2		
			16,15		
(2), A2.16, A2.15			2,2	0	
			2,3		
			2,15		
			2,16	–2.9	–5
5k, A15.3, 2.9k, A15.14, (15)			15,16	+2.9	
			15,2		–5
			15,3		
			15,14		
A14.3, 5.8, A14.13, (14), 13.75			14,15	+5.8	
		+13.75	14,3		
			14,13		
(3), 2.9, A3.4, 5.8, 13.75, A3.13			3,14		+13.75
			3,15	–2.9	–5
			3,2	–2.9	
			3,4		
			3,13		
10.1, A3.4, A13.12, 5.8, (13)			13,14	5.8	
			13,3	–5. 05	+8.75
			13,4		
			13,12		
5, 0.75, A4.5, (4), 8.75, A4.12			4,13		8.75
		–5	4,3	–0.75	
			4,5		
			4,12		
(5), 1.42, A5.6, A5.12			5,4	+1.42	
			5,6		
			5,12		

7	8	9	10	11	12	13
Unknown Axial Forces		$(7) = -(2) - (5)$ or $(8) = -(3) - (6)$	Trigonometric Results	Result of Second Equation	Resultant Axial Forces	T C
Horizontal Component	Vertical Component					
$A_{1,2}$				$A_{1,2} = 0$	$A_{1,2} = 0$	
	$A_{1,16}$	$A_{1,16} = 5$			$A_{1,16} = 5$	C
$A^h_{16,2}$	$A^v_{16,2}$	$A^v_{16,2} = 5$	$A^h_{16,2} = 2.9$		$A_{16,2} = 5.8$	T
$A_{16,15}$				$A_{16,15} = 2.9$	$A_{16,15} = 2.9$	C
$A_{2,3}$		$A_{2,3} = 2.9$			$A_{2,3} = 2.9$	T
	$A_{2,15}$			$A_{2,15} = 5$	$A_{2,15} = 5$	C
$A^h_{15,3}$	$A^v_{15,3}$	$A^v_{15,3} = 5$	$A^h_{15,3} = 2.9$		$A_{15,3} = 13.75$	T
$A_{15,14}$				$A_{15,14} = 5.8$	$A_{15,14} = 5.8$	C
	$A_{14,3}$			$A_{14,3} = 13.75$	$A_{14,3} = 13.75$	C
$A_{14,13}$		$A_{14,13} = 5.8$			$A_{14,13} = 5.8$	C
$A_{3,4}$				$A_{3,4} = 0.75$	$A_{3,4} = 0.75$	T
$A^h_{3,13}$	$A^v_{3,13}$	$A^v_{3,13} = 8.75$	$A^h_{3,13} = 5.05$		$A_{3,13} = 10.1$	T
	$A_{13,4}$	$A_{15,4} = 8.75$			$A_{13,4} = 8.75$	C
$A_{13,12}$				$A_{13,12} = -.75$	$A_{13,12} = .75$	C
$A_{4,5}$				$A_{4,5} = -1.42$	$A_{4,5} = 1.42$	C
$A^h_{4,12}$	$A^v_{4,12}$	$A^v_{4,12} = 3.75$	$A^h_{4,12} = 2.17$		$A_{4,12} = 4.3$	T
$A_{5,6}$		$A_{5,6} = 1.42$			$A_{5,6} = 1.42$	C
	$A_{5,12}$			$A_{5,12} = 0$	$A_{5,12} = 0$	

TABLE 7.1. *(Continued)*

1	2	3	4	5	6
	External Forces			Known Axial Forces	
Free Joint Diagram	Horizontal Component	Vertical Component	Bars	Horizontal Component	Vertical Component
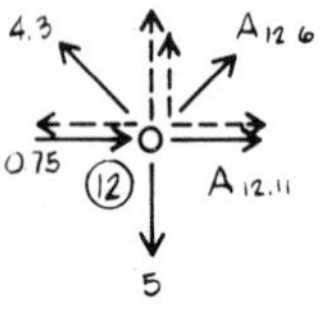 4.3, $A_{12,6}$, 0.75, (12), $A_{12,11}$, 5			12,3	+0.75	
			12,4	−2.17	+3.75
		−5	12,5		0
			12,6		
			12,11		
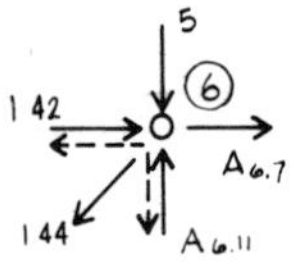 5, (6), 1.42, $A_{6,7}$, 1.44, $A_{6,11}$			6,12	−0.72	−1.25
			6,5	+1.42	
		−5	6,7		
			6,11		
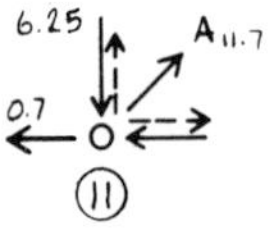 6.25, $A_{11,7}$, 0.7, (11)			11,12	−0.7	
			11,6		−6.25
			11,7		
			11,10		
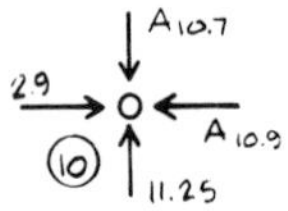 $A_{10,7}$, 2.9, $A_{10,9}$, (10), 11.25		+11.25	10,11	2.91	
			10,7		
			10,9		
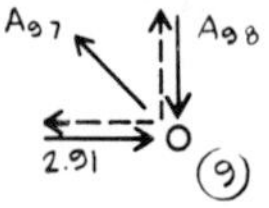 $A_{9,7}$, $A_{9,8}$, 2.91, (9)			9,10	2.91	
			9,7		
			9,8		

−(3) − (6). The items in parentheses are explained in the preceeding listing. For example, item (2) is the algebraic sum of horizontal components of external forces applied to the joint, and (5) is the algebraic sum of horizontal components of known axial forces at the

7	8	9	10	11	12	13
Unknown Axial Forces		(7) = –(2) – (5) or (8) = –(3) – (6)	Trigonometric Results	Result of Second Equation	Resultant Axial Forces	T C
Horizontal Component	Vertical Component					
$A^h_{12,6}$	$A^v_{12,6}$	$A^v_{12,6} = 1.25$	$A^h_{12,6} = 0.72$		$A_{12,6} = 1.44$	T
$A_{12,11}$				$A_{12,11} = 0.7$	$A_{12,11} = 0.7$	T
$A_{6,7}$		$A_{6,7} = -0.70$			$A_{6,7} = 0.70$	C
	$A_{6,11}$			$A_{6,11} = 6.25$	$A_{6,11} = 6.25$	C
$A^h_{11,7}$	$A^v_{11,7}$	$A^v_{11,7} = 6.25$	$A^h_{11,7} = 3.61$		$A_{11,7} = 7.2$	T
$A_{11,10}$				$A_{11,10} = 2.91$	$A_{11,10} = 2.9$	C
	$A_{10,7}$			$A_{10,7} = 11.25$	$A_{10,7} = 11.3$	C
$A_{10,9}$		$A_{10,9} = 2.91$			$A_{10,9} = 2.9$	C
$A^h_{9,7}$	$A^v_{9,7}$	$A^h_{9,7} = 2.91$	$A^v_{9,7} = 5.0$		$A_{9,7} = 5.8$	T
	$A_{9,8}$			$A_{9,8} = 5.0$	$A_{9,8} = 5$	C

joint. Positive signs in columns 9, 10, and 11 indicate that the correct sign has been assumed for an unknown bar force in the free joint diagram.

10. Trigonometric results. When a bar is not horizontal or ver-

tical, one component of its axial force is found by equilibrium. The other component is calculated according to trigonometric relations between the two components, and it is entered in this column.

11. Results of the second equilibrium equation.

12. Values of unknown axial forces at the joint. If a bar is horizontal or vertical, the entry in column 12 is the same as that in columns 9 or 11; otherwise, the bar's axial force must be calculated by trigonometric relations between the axial force and its x and y components. The magnitude of the axial forces is shown without sign in this column.

13. Axial force type: tension or compression.

We start the table by entering operations already performed and results already obtained for joints 1, 16, and 2 of the truss in Figure 7.9*a*. The remaining operations and results appear in the table without further discussion, for they proceed as usual. Using the table is, of course, purely optional, but it has the advantage of organizing and presenting the procedure in a synthetic way. Furthermore, the table can easily be programmed for automated analysis.

7.2.2. Graphic Evaluation

The method of joint equilibrium for evaluating axial forces in the bars of trusses can be employed graphically as well as numerically. The graphic procedure is known as the Maxwell method, and at each joint of the truss, it requires that the graphic condition for force equilibrium, stated in Appendix 5.3 of Chapter 5, hold: *the force polygon of a system in equilibrium must be closed.* This statement is the graphic equivalent of the two equilibrium equations

$$\Sigma Fx = 0,$$

$$\Sigma Fy = 0.$$

It is therefore necessary in the graphic as well as numerical approach to start by operating on a joint where only two bar forces are unknown and to proceed from there to a neighboring joint where unknown bar forces are reduced to only two after evaluating axial forces at the previous joint.

Loading forces, reactive forces, and known bar forces converging to a joint are drawn in the force polygon in consistent clockwise order, starting from the earliest known force, which is recognized by testing to see whether the previous force on the dial (the joint's free body diagram) is one of the two unknown bar forces. After all known forces converging to a joint have been drawn, the force polygon is closed with two sides (forces) parallel to the two unknown bar forces. These are then read with the same scale used for the known forces.

Force polygons of all the joints can be assembled in a *Maxwell diagram*, which has the advantages of the table used for the numerical procedure and, like the table, is only optional. The graphic procedure is applied for demonstration purposes to the truss in Figure 7.9*a*, so that values of axial forces obtained from the three different procedures can be proved to be identical.

Polygons of forces converging to each joint are first drawn separately and then assembled in the Maxwell diagram, using the same scale in both cases. Maxwell provided standard notations to be used with the diagram. The notations are unnecessary, however, and will be disregarded in favor of a simpler system of record keeping. The truss in Figure 7.9*a* is drawn again in Figure 7.12*a*, with its external forces sequentially labeled 0–1, 1–2, . . . , 6–7, starting with the leftmost force and moving clockwise around the truss.

The Maxwell diagram (Figure 7.12*b*) starts by drawing the closed polygon 01234567 of all external forces read in clockwise sequence around the truss. We start with joint 1 because only two unknown forces, the three- and six-o'clock forces, converge at this joint. The only known force is the 5-k, 12-o'clock force 0–1, and it is drawn as the first side of the force polygon for joint 1. We must now draw a horizontal force (the three-o'clock force) and a vertical force (six-o'clock) of such magnitude that it will close the force polygon of joint 1. For this purpose, the horizontal force must be zero and the vertical force must be 1–0 (Figure 7.13*a*). Joint 1 is at the top end of bar 1–16, the six-o'clock bar. The bar's axial force points up, thus pushing the joint. Bar 1–16 is therefore a compression bar.

In the Maxwell diagram, compressive forces are shown with solid segments, tensile forces with dashed segments.

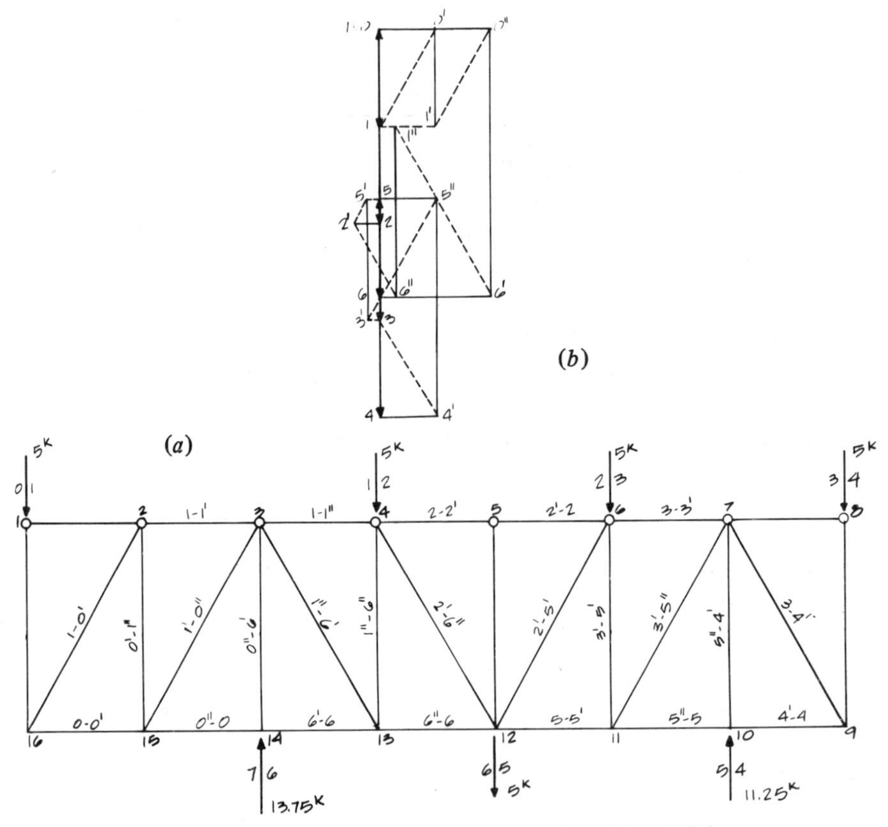

FIGURE 7.12. (*Original drawing reduced by 50%.*)

The number of unknown bar forces at joint 16 in the neighborhood of joint 1 has been reduced to only two, the one- and three-o'clock forces. The only known force is the 12-o'clock compressive axial force in bar 1–16. This force pushes joint 16 as well as joint 1 and is shown in Figure 7.13*b* as the first side of the force polygon for joint 16. We must now draw a diagonal force (one-o'clock force) and a horizontal force (three-o'clock) under the condition that the polygon of the forces converging to joint 16 must be closed (Figure 7.13*b*).

In the Maxwell diagram, the force polygon of joint 16 is the triangle 010′ whose side 1,0′ is indicated by a dashed line, since it represents a tensile axial force. Joint 16 is, indeed, at the bottom of

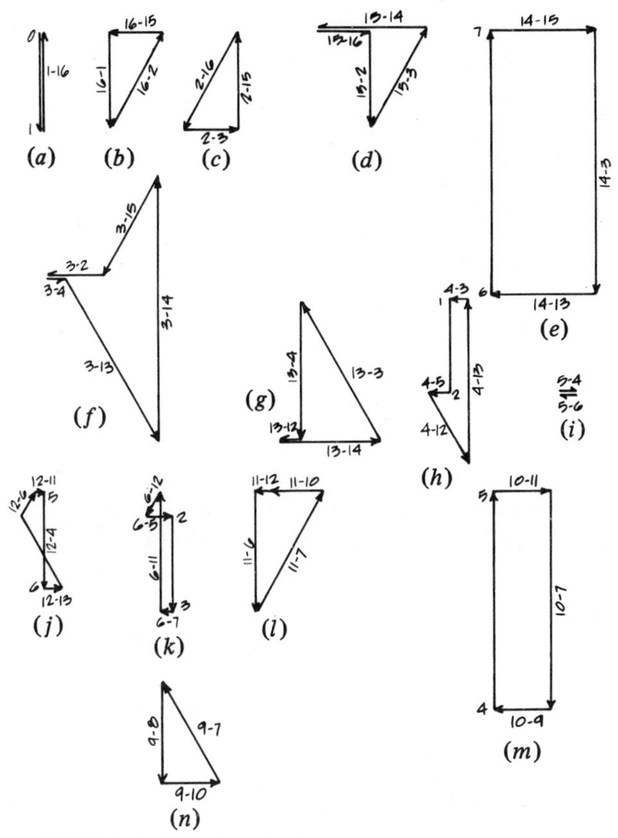

FIGURE 7.13. (*Original drawing reduced by 50%.*)

bar 2–16, and the axial force in this bar points up (Figure 7.13*b*), thus pulling joint 16. With similar reasoning, we recognize that the axial force in bar 16–15 is compressive. Joint 16 is, indeed, the left end of this bar, and the bar's axial force points left (Figure 7.13*b*), thus pushing the joint.

*In Figure 7.12*a *we write 1-0′ next to bar 16–2 and 0′-0 next to bar 16–15. These symbols immediately identify axial forces of the two bars in the Maxwell diagram. Other notations are not necessary, which is the reason for disregarding the Maxwell notations.*

The number of unknown bar forces at joint 2 has been reduced to only two, the three-o'clock and six-o'clock forces. The only known force is the axial force in bar 2–16, which appears in Figure 7.13*c* as the first side of the force polygon for joint 2. Bar 2–1 lacks axial force. The polygon is completed by a force parallel to the next bar on the dial of joint 2, which is bar 2–3, and by a force parallel to bar 2–15. Figure 7.13*c* shows that the axial force in bar 2–3 pulls joint 2 and is, therefore, tensile, while the axial force in bar 2–15 pushes the joint and is compressive. In the Maxwell diagram, the force polygon of joint 2 is the triangle 0′11′. We write 1–1′ next to bar 2–3 and the 1–0′ next to bar 2–15 in Figure 7.12*a* to identify their axial forces.

The number of unknown bar forces at joint 15 has been reduced to only two, the one- and three-o'clock forces. The earliest known force on the dial of joint 15 is the nine-o'clock force; indeed, the previous force on the dial, the three-o'clock force, is one of the two unknown bar forces. We therefore start the force polygon of joint 15 with the nine-o'clock force followed by the 12-o'clock force. The force polygon is then completed by a diagonal force parallel to bar 15–3 and by a horizontal force parallel to bar 15–14 (Figure 7.13*d*). In the Maxwell diagram, the force polygon of joint 15 is the quadrilateral 00′1′0″0. In Figure 7.12*a*, the bars converging to joint 15 are labeled accordingly for easy identification of their axial forces. The application of the joint method proceeds similarly with the other joints. All force polygons are shown on Figure 7.13, and the complete Maxwell diagram is in Figure 7.12*a*.

It can be noted from the diagram how the geometric pattern used for diagonal bars produces tension in all diagonals and how axial forces in the bars of the top and bottom chords have signs we would expect from observing the deformation of a beam with the same loads, geometries, constraints (Figure 7.14) as the truss.

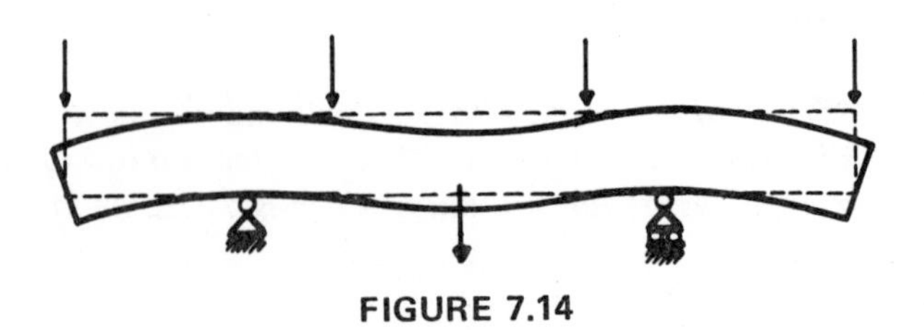

FIGURE 7.14

PROBLEMS

7.1. Using the Ritter section method, find the axial forces in the bars of the second panel from the left end (section S) (Figure 7.15).

Solution. Step 2 provides reactions of the external constraints. The equation of moment equilibrium around the hinge is

$$\Sigma M = 0 = 1(8.66) - 6(5) - 5(10) - 8(15) + 20V_r + 1(8.66),$$

from which

$$V_r = \frac{1}{20}(-8.66 + 30 + 50 + 120 - 8.66) = 9.13 \text{ k.}$$

The equation of vertical equilibrium yields

$$V_l = 9.87 \text{ k.}$$

The equation of horizontal equilibrium yields the horizontal reaction of the hinge

$$H = 2 \text{ k.}$$

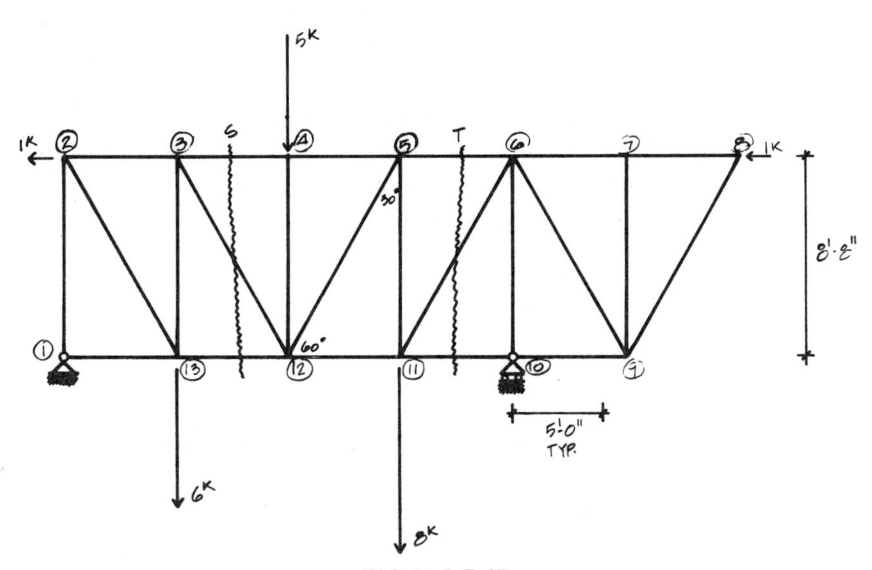

FIGURE 7.15

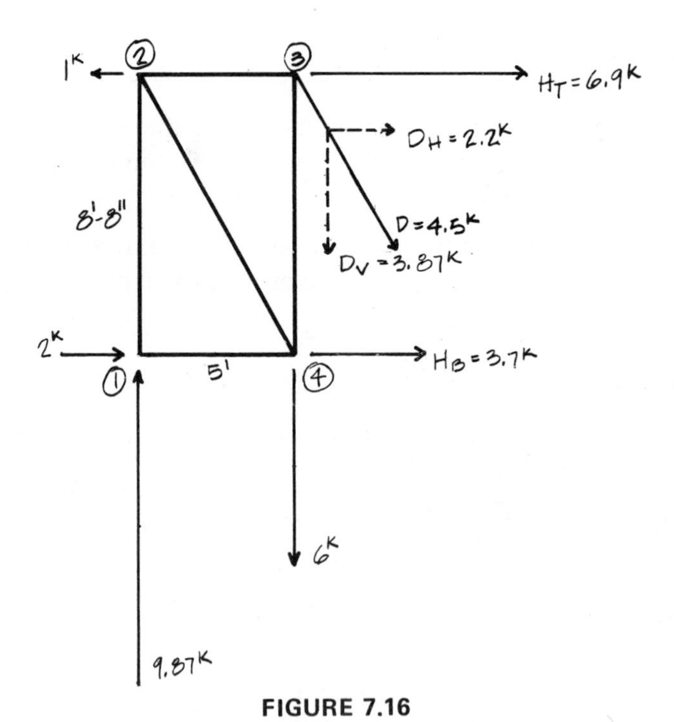

FIGURE 7.16

The free body diagram of the part to the left of the Ritter section is shown in Figure 7.16 for Step 3, which is the evaluation of axial forces. The equilibrium of vertical forces applied to the free body in Figure 7.16 is stated in equation form by

$$\Sigma F_y = 0 = 9.87 - 6 - D_v,$$

$$D_v = 3.87 \text{ k.}$$

From the geometries of the typical panel, we obtain

$$D_h = D_v \tan 30° = \frac{3.87}{(3)^{1/2}} = 2.23 \text{ k,}$$

then

$$D = \frac{D_v}{\cos 30°} = 3.81 \frac{(2)}{(3)^{1/2}} = 4.46 \text{ k.}$$

The equilibrium of moments around point C is stated by

$$\Sigma M_C = 0 = -9.87(5) + 1(8.66) - 8.66H_t - 2.23(8.66),$$

from which

$$H_t = \frac{-9.87(5)}{8.66} + 1 - 2.23 = -6.93 \text{ k}.$$

The negative result requires reversing the sign assumed for H_t. The equation of horizontal equilibrium with the correct sign of H_t states

$$\Sigma F_x = 0 = 2 - 1 - 6.93 + 2.23 + H_b,$$

from which

$$H_b = 3.7 \text{ k}.$$

7.2. Find the same results as in Problem 7.1 by solving equations of equilibrium for the free body on the right side of the same Ritter section.

7.3. Using the Ritter section method, find the axial forces in the three bars of the truss cut by the section S in Figure 7.22. Check the answers with the graphic solution in Figure 7.23.

7.4. Using the Ritter section method, find the three axial forces in bars cut by the section S in Figure 7.24*a* and check the answers with the graphic solution in Figure 7.24*b*.

7.5. Using the Ritter section method, find the axial forces in the bars of the section shown in Figure 7.25*a*. Check the answers with the graphic solution in Figure 7.25*b*.

7.6. Find the axial forces in the bars of the truss in Figure 7.15 by the numerical joint equilibrium method. Compile operations and results in tabular form.

7.7. Axial forces in the bars of the truss in Figure 7.15 are evaluated in Figure 7.17 by the graphic joint method. Assemble individual force polygons of each joint in the Maxwell diagram.

7.8. Perform the tasks in Problem 7.6 for the truss in Figure 7.18.

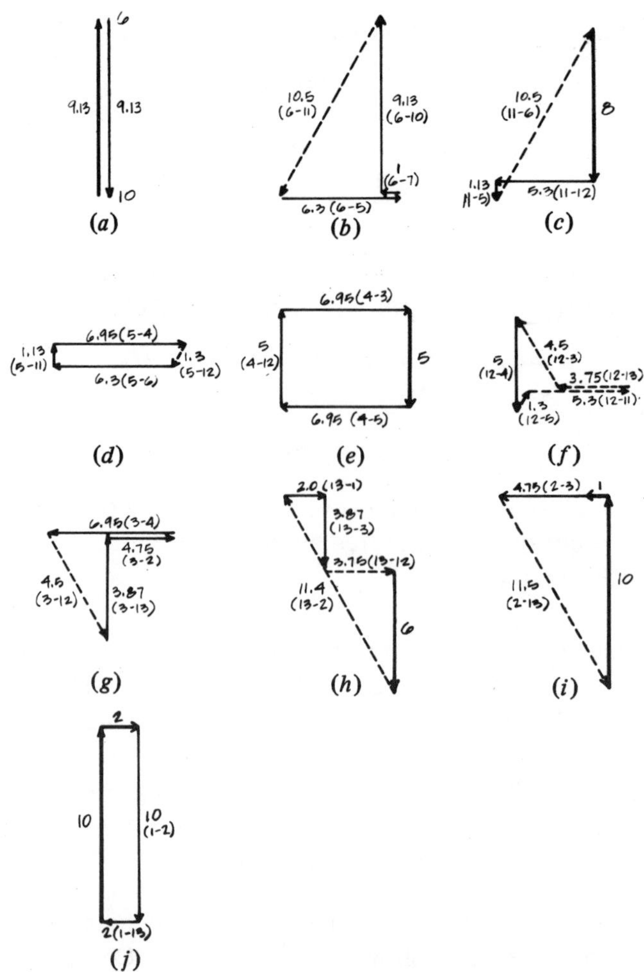

FIGURE 7.17. (*Original drawing reduced by 50%.*)

7.9. Perform the task in Problem 7.7 for the truss in Figure 7.18 (Figure 7.19).

7.10. Axial forces in the bars of the left part of the truss are shown in the Maxwell diagram. Repeat the graphic joint method, but draw separate force polygons for the equilibrium of each joint (Figures 7.20 and 7.21).

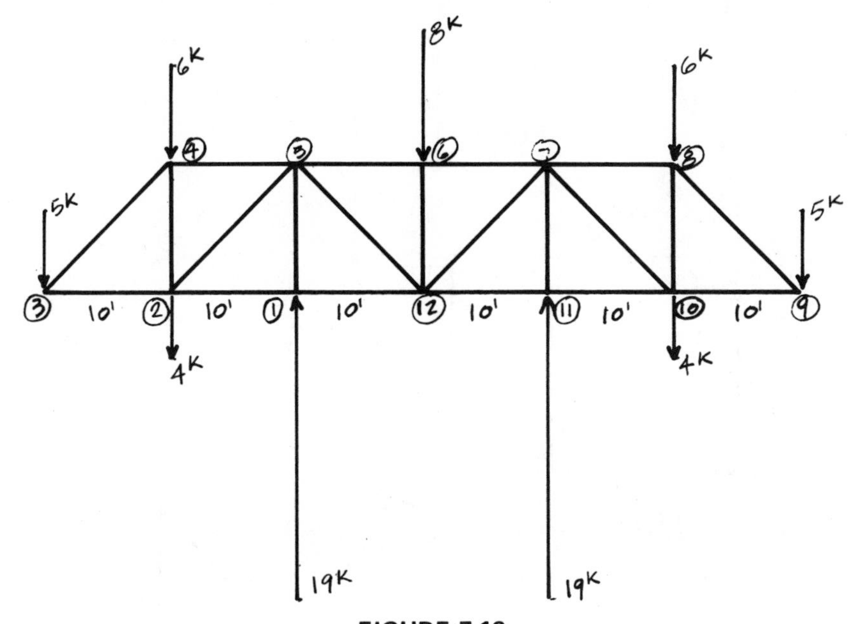

FIGURE 7.18

7.11. Perform the task in Problem 7.6 for the symmetric truss shown in Figure 7.22 with its loads and reactions. Check the results with the graphic solution in Figure 7.23.

7.12. In a Maxwell diagram, assemble the separate force polygons drawn in Figure 7.23 for each joint of the truss in Figure 7.22.

7.13. Perform the task in Problem 7.6 for the truss in Figure 7.24*a*.

7.14. Complete the evaluation of axial forces in the bars of the truss in Figure 7.24 by the graphic joint equilibrium method.

7.15. Draw a Maxwell diagram with the axial forces of the bars of the truss in Figure 7.24.

7.16. Perform the task in Problem 7.6 for the truss shown in Figure 7.25*a* with its loads and reactions. Check the results with the graphic solution in Figure 7.25*b*.

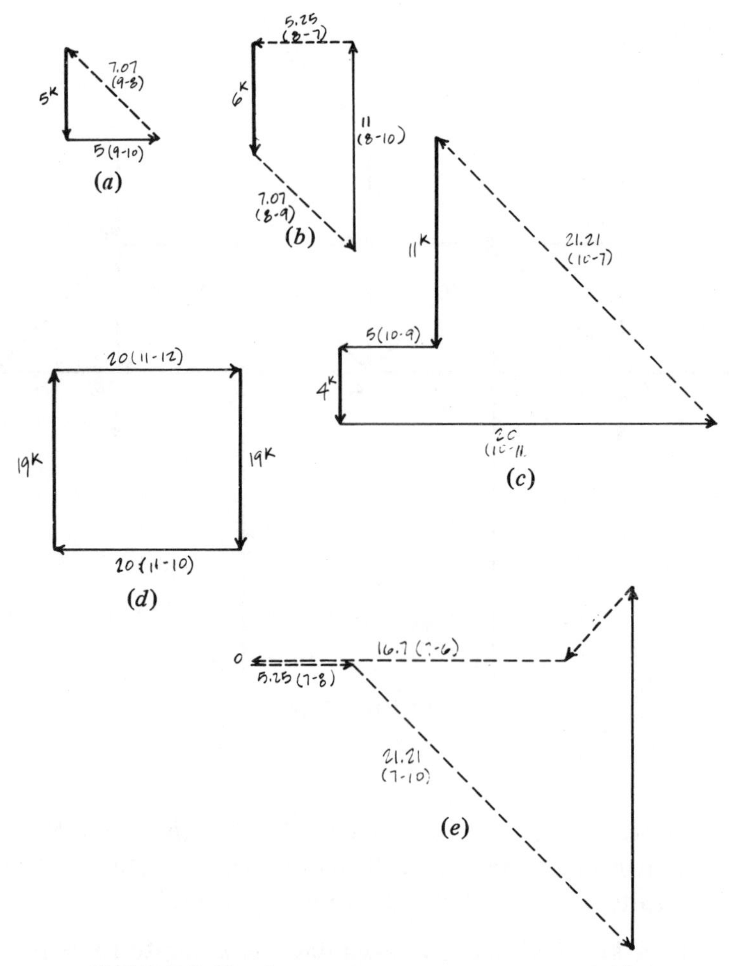

FIGURE 7.19. (*Original drawing reduced by 50%.*)

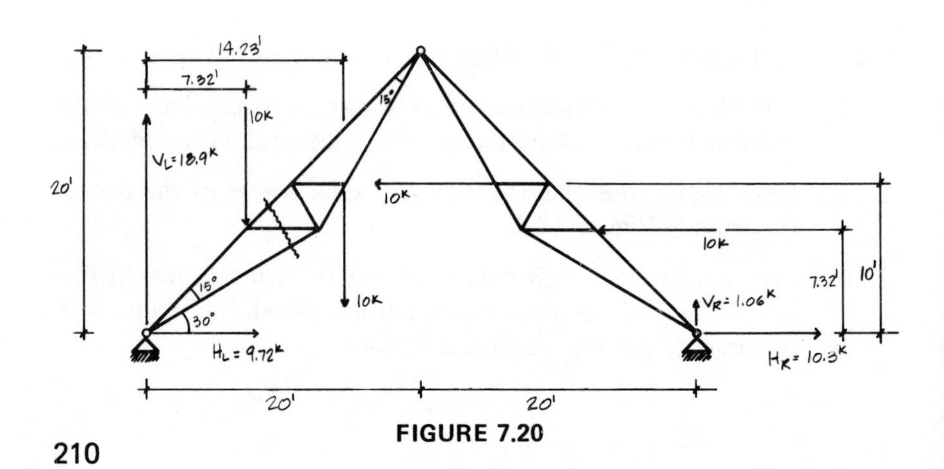

FIGURE 7.20

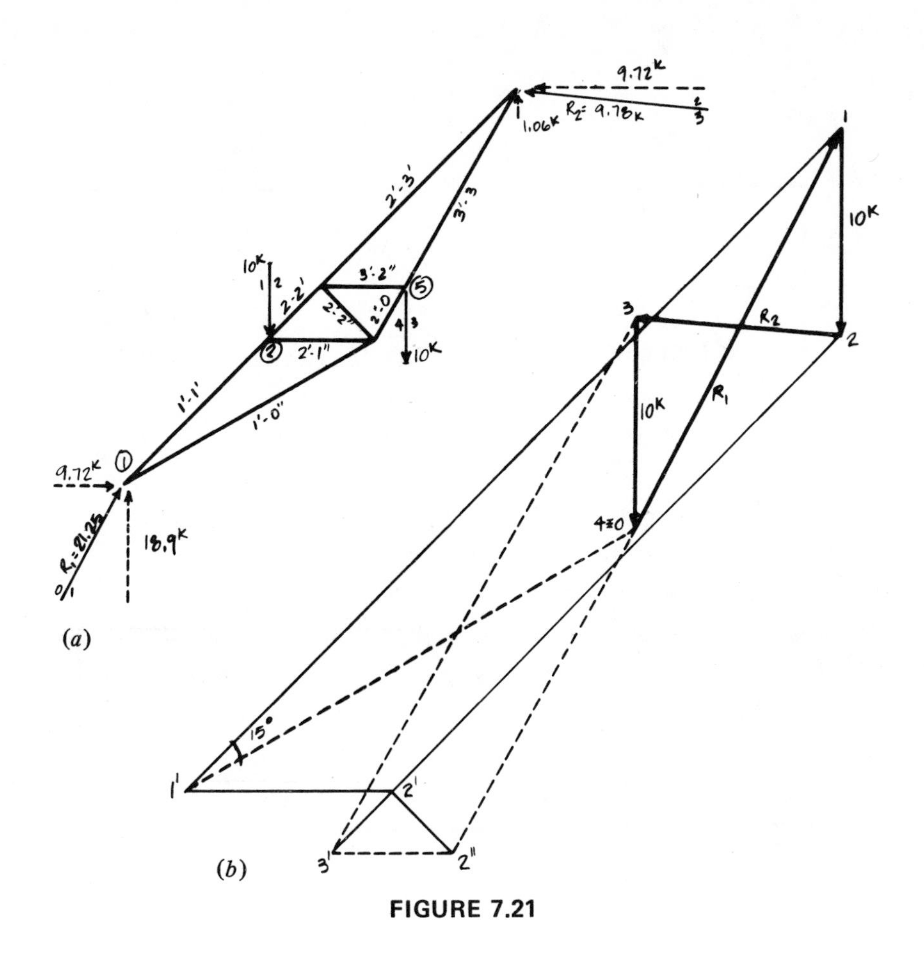

FIGURE 7.21

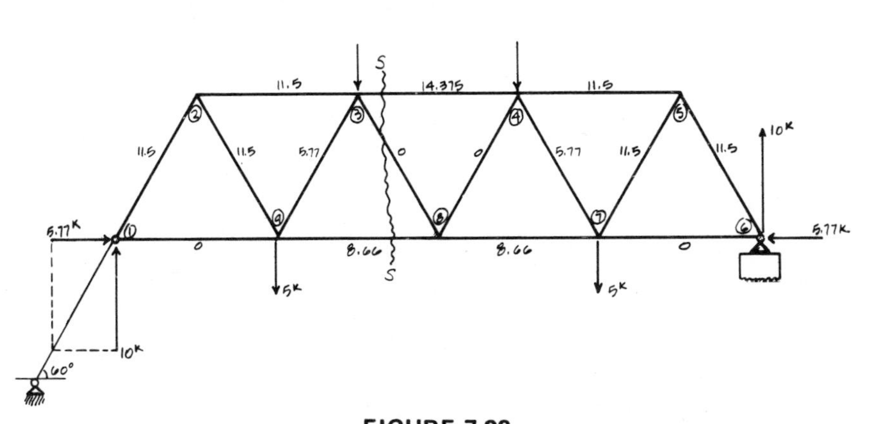

FIGURE 7.22

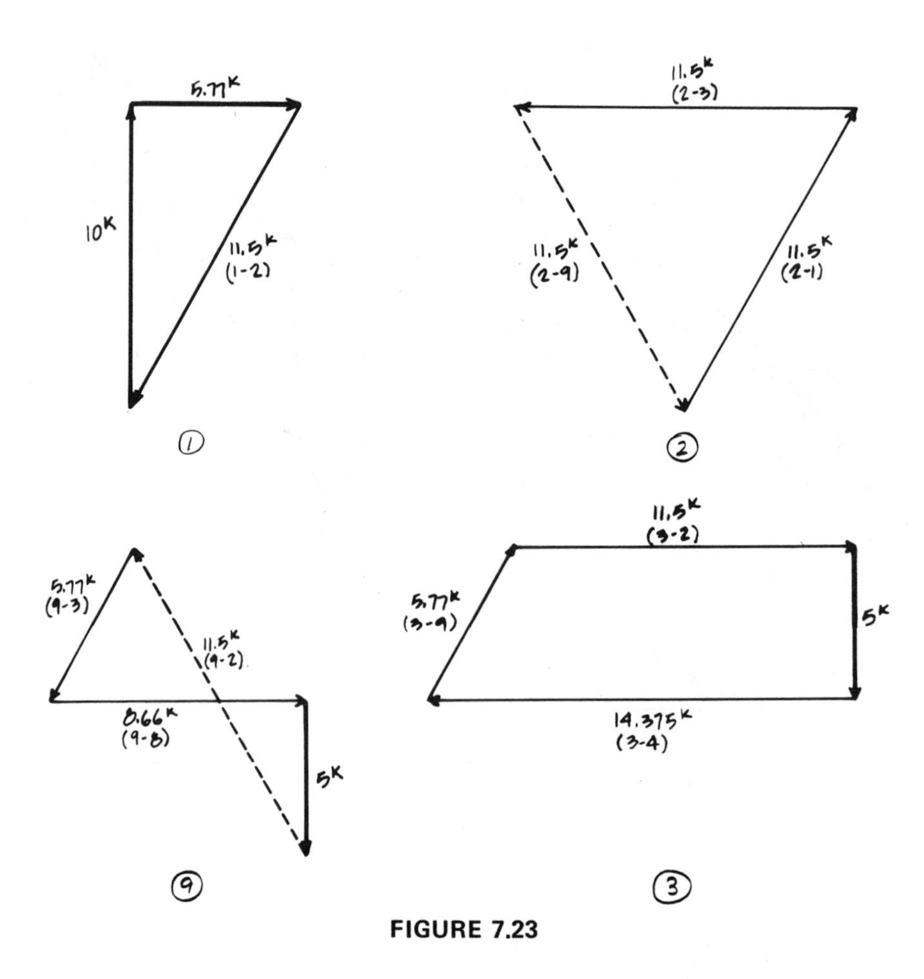

5.77K
10K
11.5K
(1-2)
1
11.5K
(2-3)
11.5K
(2-9)
11.5K
(2-1)
2
5.77K
(9-3)
11.5K
(9-2)
8.66K
(9-8)
5K
9
11.5K
(3-2)
5.77K
(3-9)
5K
14.375K
(3-4)
3

FIGURE 7.23

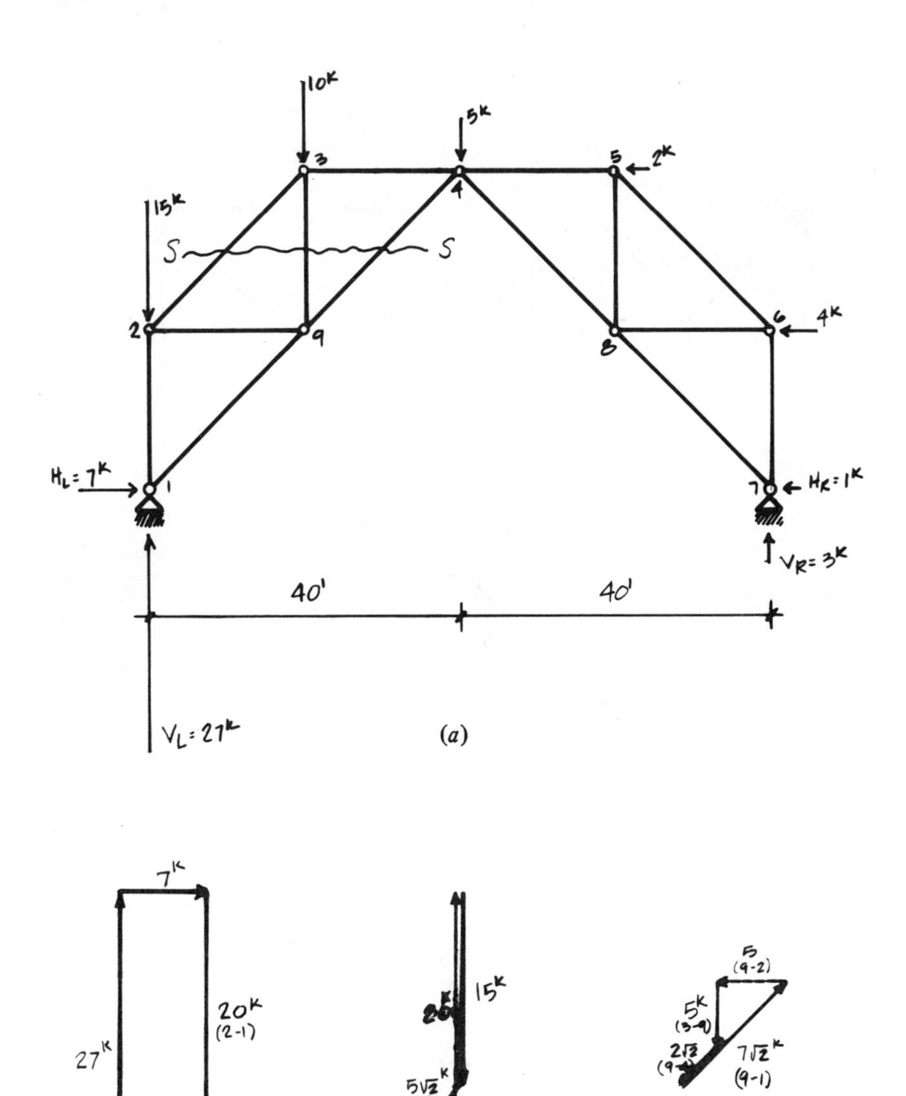

(a)

(b)

FIGURE 7.24

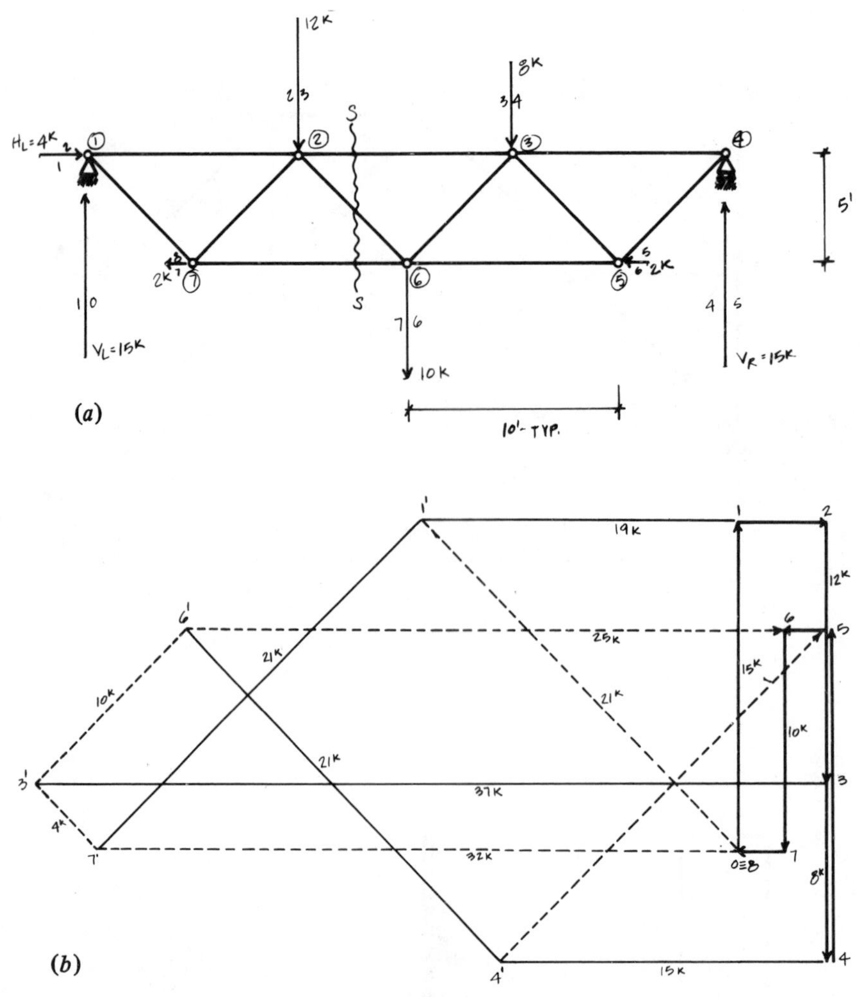
12K
8K
HL=4K
VL=15K
VR=15K
10K
2K
2K
5'
10'- TYP.
(a)
19K
12K
25K
21K
21K
21K
21K
10K
15K
10K
37K
4K
32K
8K
15K
(b)

FIGURE 7.25

7.17. Using the Ritter method, find the axial forces in the three bars of the truss cut in Figure 7.25*a* by section *S*. Check the answers with the graphic solution in Figure 7.25*b*.

7.18. Evaluate axial forces in the bars of the truss in Figure 7.25*a* with the graphic joint method. Draw separate force polygons for the equilibrium of each joint, and check the answers in the Maxwell diagram in Figure 7.25*b*.

The Jefferson National Expansion Memorial, St. Louis, Missouri. Architect: Eero Saarinen. Structural Engineers: Severud-Perrone-Szegezdy-Sturm, New York. Photograph courtesy of Severud-Perrone-Szegezdy-Sturm.

EIGHT

ARCHES

Thousands of years ago, some of the structural elements that are now available were not known. Some of them could not be built for lack of necessary structural materials or techniques of connection and fabrication. Beams could be made only of timbers, and they could extend only over modest spans, since wood is not a very strong material. Horizontal stone elements carrying gravity loads were also limited for use over modest spans, since all brittle materials have the capacity to resist shear forces associated with flexural members but little capacity for carrying the tension caused by bending in the top or bottom fibers of beams. Cables were known only as ropes made of natural fibers, and they could be used only for temporary structures like tents and small river bridges. Trusses could be built only to the extent that timber bars connected by nails would permit.

The Romans who ruled a colonial empire nearly as large as the known world at that time, had to build an unprecedented number of large structures necessary for their administration. These structures included roadway bridges, aqueducts, amphitheaters, baths, temples, assembly halls, administrative buildings, and so on. Such buildings required rigid structures, made of strong and durable materials, covering considerable spans.

Arches made with stones and mortar, materials then available, were the natural answer to those needs. Indeed, arches are rigid structures and, due to their efficient behavior, can cover very large spans. These arches include some Roman bridges that are still standing today over the Tiber River, proving the durability of the materials and the strength of those ancient structures. Building arches with bricks and mortar was a simple although time consuming practice suitable in an age of plentiful if unspecialized labor. Even a few years ago, in regions still under the strong influence of Roman culture, it was possible to find masons skilled in building daring and gracefully rampant arches supporting flights of stairs or sloping roofs.

The efficiency of arches is related to their capacity to transfer loads to supports by developing axial forces rather than shear forces and bending moments, a feature also shared by cables and trusses. This capacity is, of course, greater or lesser depending on the degree to which some ideal conditions are fulfilled. Specifically, an ideal arch, one where shear and bending are totally absent, is realized by turning a cable upside down, which changes the axial forces from tension to compression without introducing shear and bending. Such arches do not exist. Indeed, a rigid structure does not have a cable's capacity to move into a geometric configuration that eliminates distances between axial forces and the centers of the cross sections to which they apply. Eccentric axial forces produce moments, which are inevitably accompanied by shear.

Under the best hypothetical conditions, an arch can, therefore, be free of bending under only one load condition, that is, under the action of dead loads. As the load condition changes, the funicular polygon of the loads (the reversed cable shape) also changes. The rigid arch preserves its original shape; therefore, distances materialize between the arch and axial forces that are colinear with the sides of the new string polygon. Bending moments in the arch are proportional to these distances. The deviation of the string polygon from the geometric axis of the arch can, indeed, be used as the diagram of bending moments with a variable scale that coincides at each cross section with the local axial force.

An example is given in Figure 8.1, where arch ABC is an A frame loaded, for simplicity, by a concentrated load F on only one side. The string polygon of the load (the cable shape in reverse) is the bilateral ADC. The reaction R_C of constraint C must, in fact, be in the direction of the unloaded link BC, and the reaction of constraint A must be a force through hinge A equal and opposite to the resultant of F and R_C. Since a cable carrying a concentrated load must have a bilateral shape with the direction of R_A at the end A and the direction of R_C at the end C, polygon ADC identifies the reversed cable shape (string polygon) by satisfying those conditions.

The points of the arch along AE have the typical distance d from R_A, shown in Figure 8.1. Therefore, at a given point P of AE, the

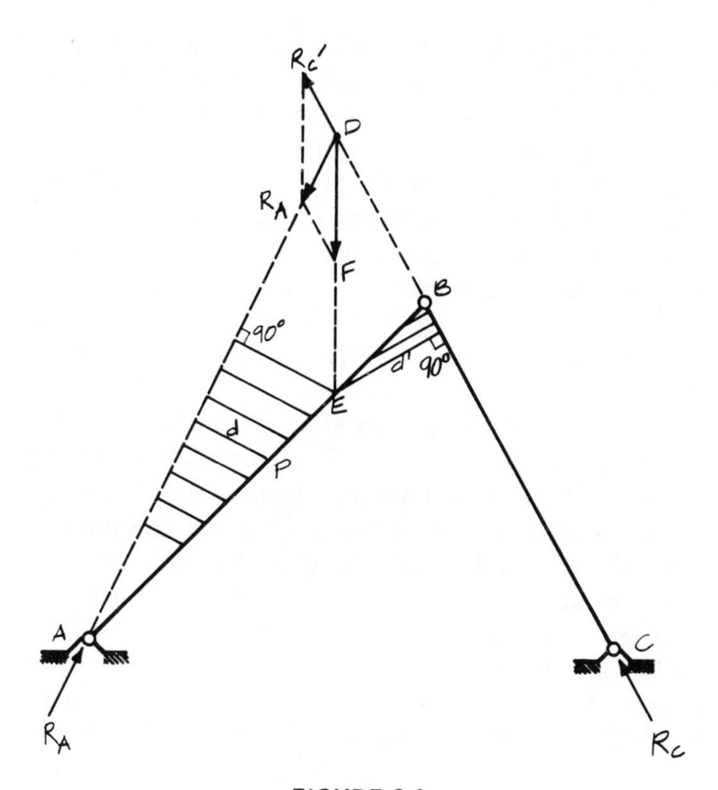

FIGURE 8.1

bending moment is the product $R_A\,d$ of R_A with the typical distance d. In a scale of moments S_M numerically equal to R_A, d can represent the bending moment at P. The moment would, in this case, be

$$S_M d = R_A\,d.$$

Similarly, the typical moment along EB can be represented by the typical distance d' in a scale numerically equal to R_C. This reaction coincides with the internal force at all points of the arch from B to E, and its product with the typical distance d' is, therefore, the typical bending moment along BE.

To summarize, the arch is bending free along segment BC because all of its points from B to C belong to the reversed cable shape. Segment AB of the arch is subject to bending moments proportional to the distance, such as d, between the reversed cable shape and the points of AE. Segment EB of the arch is subject to bending moments proportional to the distance, such as d', between the reversed cable shape and the points of EB. For optimal structural efficiency of an arch, the geometric axis should be given a shape as close as possible to the string polygon of the prevalent load condition.

The choice of external and internal constraints is another important design decision. In the introduction to Chapter 6, we said that it was not necessary to specify constraint conditions for a cable because every point, including the ends, behaves like a hinge. In the case of arches, external and internal constraints can be used occasionally to move the string polygon of the loads close to the geometric axis of the arch—at least at single points. An arch fixed at both ends is threefold redundant. Therefore, any combination of moment releases (hinges) and shear releases (slides) not exceeding a total number of three can be provided along the geometric axis and at its ends without affecting the stability of the arch.

Since the bending moment must be zero at a hinge, the string polygon of the loads touches the geometric axis of the arch at the hinged sections. Since the moment reaches a maximum where the shear is zero, a shear release can condition the string polygon to reapproach the axis of the arch on both sides of the slide. The judicious placement of these releases with respect to the positions of loads can result in improved structural efficiency.

For example, the bending moments on the semicircular, statically indeterminate arch in Figure 8.2*a* are given by the ordinates of the half sine wave

$$M = \frac{Pr}{2}(1 - \sin\theta).$$

If the cross section of the arch is rectangular, the horizontal reaction of the hinges is approximately P/π, and the string polygon of load P is as shown in Figure 8.2*a*. Placing a hinge at midspan means that the string polygon of load P must touch the axis of the arch at midspan (Figure 8.2*b*), since the hinge makes the midspan moment vanish.

The ordinates of the moment diagram are those of the wave

$$M = \frac{Pr}{2}(1 - \cos\omega - \sin\omega).$$

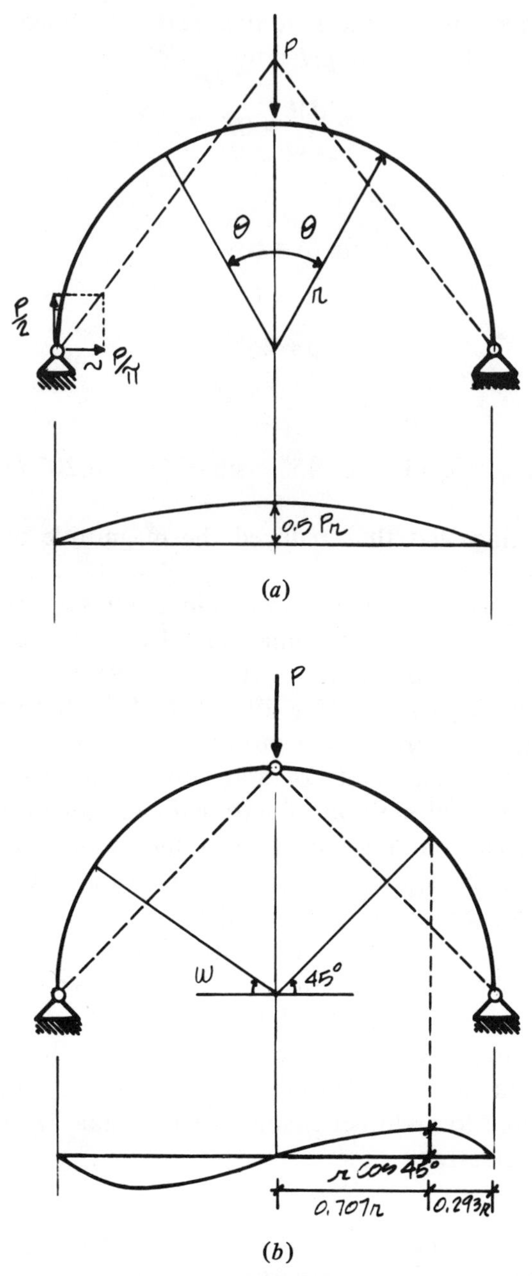

FIGURE 8.2

The maximum moment occurs at the section inclined on the horizontal plane with the angle ω given by

$$\frac{dM}{d\omega} = 0,$$

from which

$$\sin \omega = \cos \omega$$

and

$$\omega = 45°.$$

Then,

$$M_{max} = \frac{Pr}{2}(1 - \cos 45° - \sin 45°) = -0.207\,Pr.$$

The midspan hinge has thus reduced the magnitude of M_{max} from 0.5 to 0.207 Pr.

A temporary use of releases can eliminate stresses induced in the early life of a statically indeterminate arch by creep and shrinkage of the material and by foundation settlements. Releases can be eliminated by concreting and welding after distortions have occurred, so that live loads are carried by a more rigid and stronger structure. Reactions of the constraints of statically determinate arches are obtained by the usual method of setting up and solving equations of equilibrium. Each combination of shear and moment releases along the axis of the arch requires a different formulation of the equilibrium conditions, and each one is investigated separately.

Figure 8.3 shows all possible constraint conditions of statically determinate arches. The three-hinge arch (Case A) is the most frequently constructed statically determinate arch. The other cases are seldom, if ever, found, but we consider them here for practice in evaluating reactions and for exploring options in structural design that are available to architect-engineers whose goal is to design with efficiency and creativity.

In the statement of equilibrium conditions that give reactions of the constraints, the Cartesian frame of reference must have the x-axis extending from one terminal pivot of rotation to the other in all cases.

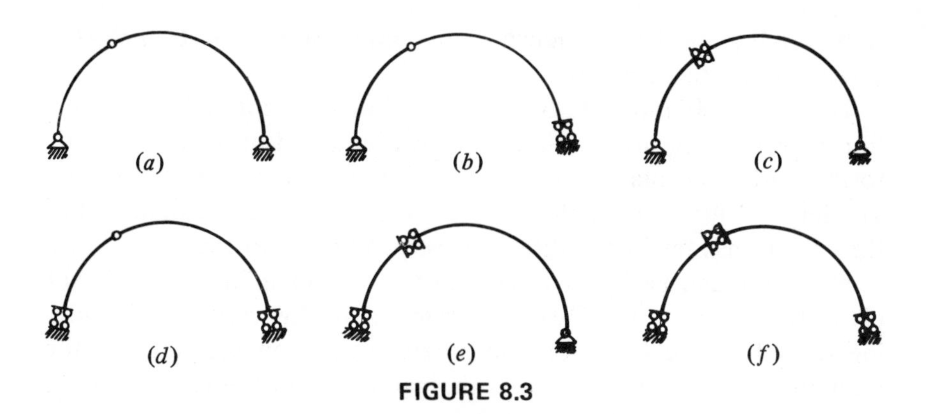

FIGURE 8.3

Such a Cartesian frame simplifies the three equations

$$\Sigma F_x = 0,$$

$$\Sigma F_y = 0,$$

$$\Sigma M = 0.$$

For example, in the arch in Figure 8.4, the left-end hinge is one terminal pivot of rotation, and the right-end hinge is the other terminal pivot of rotation. The x-axis of the Cartesian frame is, therefore, the horizontal line extending from the left hinge to the right hinge.

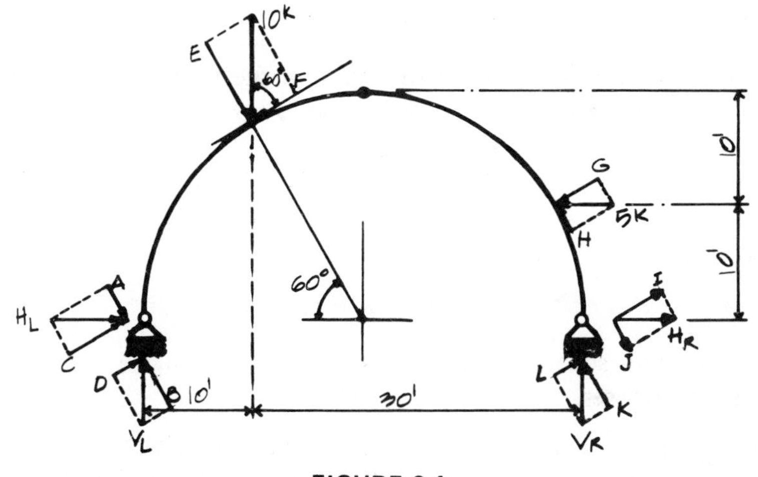

FIGURE 8.4

If the arch is neither symmetric nor symmetrically constrained, the x-axis will not be horizontal.

The arch in Figure 8.6, for example, has a horizontal left-end section hinged to the foundation. The right-end section is inclined 30° to the horizontal plane, and it is slide-connected to the foundation. The left-end hinge is one terminal pivot of rotation. The right end of the arch translates in the direction perpendicular to the links. Since a translation can be defined as a rotation around an infinitely distant pivot, the x-axis of the Cartesian frame is, in this case, the line joining the left hinge to a point infinitely distant on the links of the right-end section. In short, x is the parallel from the left-end hinge to the links of the right-end section.

Case A

The equations of equilibrium are written in this order:

First equation: equilibrium of moments about one end to obtain the component of the reaction parallel to y at the other end. Thus, for the arch in Figure 8.4, choosing the left hinge for pivot, we state

$$\Sigma M = 0 = -10(10) + 5(10) + 40V_R,$$

from which

$$V_R = 1.25 \text{ k}.$$

Second equation: equilibirum of forces in the y direction to obtain the y component of the other end reaction. For the arch in Figure 8.4, the second equation is

$$\Sigma F_y = 0 = V_L - 10 + V_R,$$

from which

$$V_L = 10 - V_R = 8.75 \text{ k}.$$

The first and second equations are written similarly for beams, cables, and trusses.

Third equation: The bending moment vanishes at the intermediate hinge. This equation is similar to one used to obtain the horizontal reaction of a cable, and it indeed yields the horizontal reaction at one end of the arch. Calculating the bending moment from the left, the third equation for the arch in Figure 8.4 is

$$M = 0 = -20V_L + 20H_L + 10(10),$$

from which

$$H_L = \frac{1}{20}(20 \cdot 8.75 - 100) = 3.75 \text{ k}.$$

Fourth equation: equilibrium of forces in the x direction, to obtain the x component of the reaction at the other end.

$$\Sigma F_x = 0 = H_L - 5 - H_R.$$

from which

$$H_R = H_L - 5 = -1.25 \text{ k}.$$

The minus sign in the result alerts us to an incorrect assumption about the sign of H_R, which must, therefore, be reversed.

A three-hinge arch has, in summary, four reactions—as many as a cable has. Both structures, indeed, have hinged ends that develop, by reason of the sag and rise, reactive forces in both the y and x directions regardless of the direction of the loads.

With external forces completely defined, evaluating internal forces at any section S proceeds as usual by summing all external forces of the same kind applied on one of the two parts of the arch separated by section S. The shear force on a typical section of the arch has the direction of the local radius of curvature of the geometric axis.

The axial force is parallel (under ideal conditions, it is colinear) to the local tangent to the geometric axis. The loads and reactions on one side of a typical section must, therefore, be decomposed in the direction of the radius and the tangent. Components parallel to the radius contribute to the shear; components parallel to the tangent contribute to the axial force.

As an example, shear and axial forces are found at the section where the 10-k vertical load is applied. Working with the part of the arch on the left of this section, we must decompose V_L and H_L in forces A and B parallel to the radius and C and D parallel to the tangent. Without including the components of the 10-k load, the shear is thus,

$$B - A = V_L \cos 30° - H_L \cos 60° = 8.75 \frac{(3)^{1/2}}{2} - \frac{3.75}{2} = 5.7 \text{ k}.$$

The axial force is

$$C + D = H_L \sin 60° + V_L \sin 30° = 3.75 \frac{(3)^{1/2}}{2} + \frac{8.75}{2} = 7.62 \text{ k.}$$

At the point of application of the 10-k load, values of the shear and axial forces have discontinuities equal, respectively, to the radial and tangential components of 10 k. Thus, including the load, the new values of the internal forces are

$$\text{Shear} = B - A - E = 5.7 - 10 \cos 30° = -2.96 \text{ k},$$

$$\text{Axial Force} = C + D - F = 7.62 - 10 \sin 30° = 2.62 \text{ k.}$$

These values can also be calculated by working on the part of the arch on the right side of the 10-k load, in which case, we obtain (Figure 8.4)

$$\text{Shear} = K - J + H = V_R \sin 60° - H_R \cos 60° + 5 \cos 60°$$
$$= 2.96 \text{ k},$$

$$\text{Axial Force} = L + I - G = V_R \cos 60° - H_R \sin 60° + 5 \sin 60°$$
$$= -2.62 \text{ k.}$$

The bending moment is calculated as usual by summing the products of external forces with their arms on the left or right side of the 10 k load. Working on the left part, we obtain, for this case,

$$M = -10V_L + (20 \sin 60°) H_L = -87.5 + 17.32(3.75) = -22.5 \text{ k ft.}$$

Working on the right part, we obtain

$$M = 30V_R + (20 \sin 60°) H_R - (20 \sin 60° - 10)5 = 22.5 \text{ k ft.}$$

Components of the external forces and arms of the external forces can be measured from scale drawings of the arch and its loads and reactions when calculations need not be extremely precise.

The graphic solution of Case A is obtained as shown in Figures 8.5*a* and *b*.

When the segment *MR* of the arch is free of loads, it works as a link or pendulum reacting with a force R_{R1} along the line *RM*. Since reactions R_{L1} and R_{R1} are the components of a force equal, opposite, and colinear with the 10-k load, the line of action of R_{L1} is

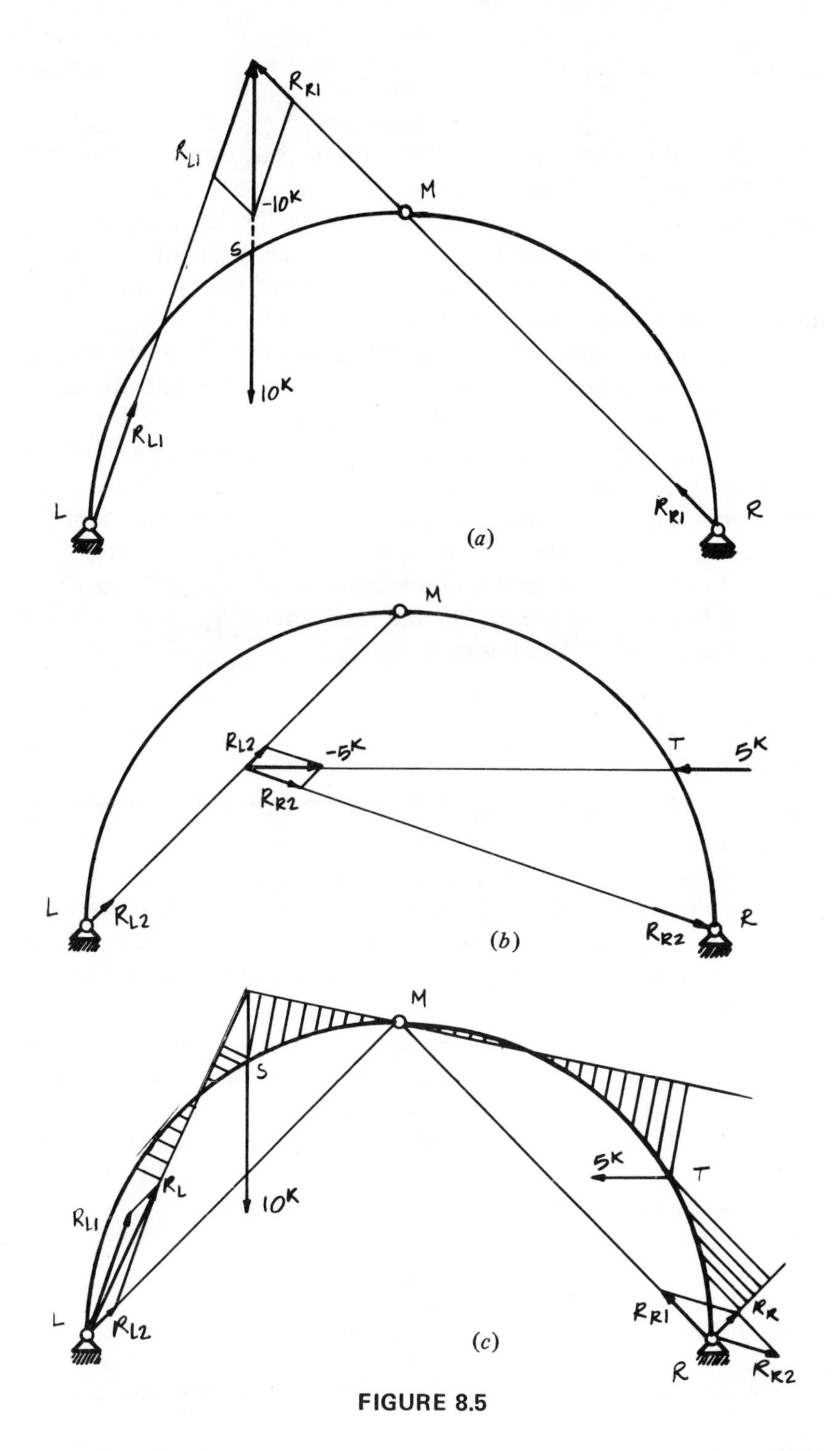

R_{R1}
R_{L1}
-10^K
M
S
10^K
R_{L1}
L
R_{R1}
R
(a)
M
R_{L2}
-5^K
T
5^K
R_{R2}
L
R_{L2}
R_{R2}
R
(b)
M
S
5^K
T
R_L
10^K
R_{L1}
L
R_{L2}
R_{R1}
R_R
R
R_{R2}
(c)

FIGURE 8.5

drawn from the left hinge L to the point of concurrence of the 10-k load with R_{R1}. The magnitudes of R_{L1} and R_{R1} are obtained by decomposing the equilibrant of the load along the lines of action of the reactions by the parallelogram rule. If the segment LM of the arch carries more than one load, it is necessary first to find the resultant of the various loads.

Using a similar procedure, we find the reactions R_{L2} and R_{R2} of the external hinges to loads on segment MR of the arch. By superposition, R_L is the resultant of R_{L1} and R_{L2}. R_R is the resultant of R_{R1} and R_{R2} (Figure 8.5*c*). The moment diagram is drawn in Figure 8.5*c* with the same procedure used for the diagram of Figure 8.1. The diagram has a variable scale which at a given section is numerically equal to the magnitude of the local axial force. Therefore, from point L to point S the scale is numerically equal to R_L. From S to T the scale is numerically equal to the resultant of R_L with the 10 k load. From T to R the scale is given by R_R.

Case B

Figure 8.6 shows a statically determinate arch with the left end hinged and the right end sliding on the foundation. A moment release (hinge) connects the left circular quadrant with a semiparabolic axis. The x axis of the Cartesian frame connects the left-end pivot to the right-end pivot which is infinitely distant in the direction of the links. Thus x is parallel to the links drawn from the left-end hinge.

With traditional notations, we label V_L the x component of the left-end reaction, and V_R the right-end reaction, which is parallel to x and therefore does not have a y component. H_L is the y component of the hinge's reaction. M_R is the reactive moment of the slide. The four equations of equilibrium that yield V_L, H_L, V_R, and M_R are written in the following order: First equation

$$\Sigma F_y = 0 = H_L - B - C,$$

which yields

$$H_L = B + C = 10 \cos 60° + 5 \cos 30° = 5 + 4.33 = 9.33 \text{ k}.$$

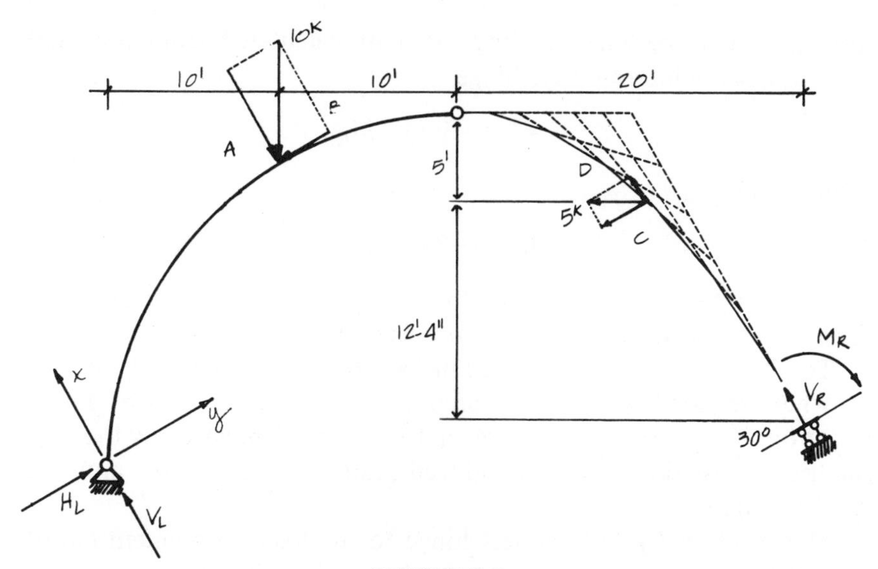

FIGURE 8.6

Second equation: The bending moment calculated from the left vanishes at the intermediate hinge.

$$M = 0 = -V_L(20\cos 30° + 20\sin 30°) + H_L(20\cos 30° - 20\sin 30°) + 10(10).$$

Dividing each term by 20, we obtain

$$V_L(0.866 + 0.5) = 5 + 9.33(0.866 - 0.5),$$

from which

$$V_L = \frac{5 + 9.33(0.366)}{1.366} = 6.16\text{ k}.$$

The arms of V_L and H_L around the pivot can also be satisfactorily approximated from a scale drawing of the arch. Third equation

$$\Sigma F_x = 0 = V_L - A + D + V_R,$$

which yields

$$V_R = A - D - V_L = 10\sin 60° - 5\cos 60° - 6.16$$
$$= 8.66 - 2.5 - 6.16 = 0\text{ k}.$$

Fourth equation: The bending moment calculated from the right vanishes at the intermediate hinge

$$M = 0 = -5(5) - M_R,$$

from which

$$M_R = -25 \text{ k-ft.}$$

M_R is counterclockwise rather than clockwise, as we had assumed. The evaluation of internal forces proceeds as in Case A.

To obtain the graphic solution in Case B, we reason as follows:

When segment MR of the arch (Figure 8.7*a*) is free of loads, reaction V_R of the links (parallel to the links) must cross the geometric axis of the arch at the hinged section M, so that the moment vanishes at M.

The reaction R_{L1} of the left hinge to the load on segment LM of the arch concurs with the 10-k load and R_{R1}, because these three forces are in equilibrium. The magnitude of R_{L1} and R_{R1} is thus obtained by decomposing the equilibrant of the 10-k load by the parallelogram rule along the lines of action of R_{L1} and R_{R1}.

When segment LM of the arch is unloaded, reaction R_{L2} of the left hinge has the direction of the dotted line LM. Reaction R_{R2} of the links concurs with R_{L2} and the 5-k load on segment MR of the arch. The magnitude of R_{L2} and R_{R2} is obtained by the parallelogram rule, as shown in Figure 8.7*b*. By superposition, reaction R_L is the vectorial sum of R_{L1} and R_{L2}; it is also obtained by the parallelogram rule. Reaction R_R is the algebraic sum of R_{R1} and R_{R2}. Reactive moment M_R is given by the sum of $e_1 R_{R1}$ and $e_2 R_{R2}$. The moment diagram based on the geometric axis of the arch is obtained as in Case A.

Case C

Figure 8.8 shows a semicircular arch with hinged ends and an intermediate shear release along its axis. The four equations of equilibrium that yield V_L, V_R, H_L, and H_R are written in the following order. First equation: The right (or left) end of the arch is bending free due to the hinged connection

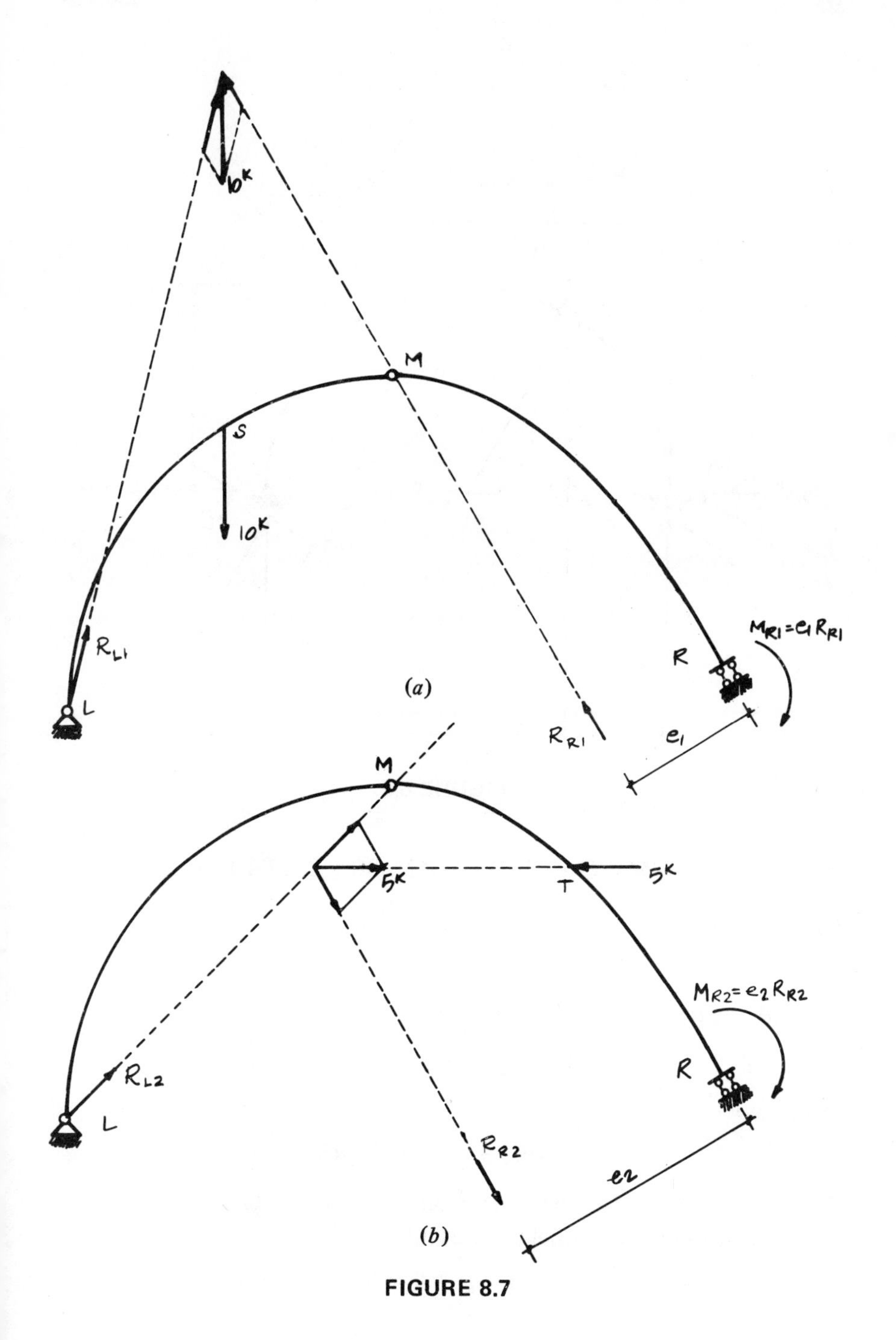

10ᴷ
M
S
10ᴷ
R_{L1}
$M_{R1} = e_1 R_{R1}$
R
(a)
L
R_{R1}
e_1
M
5ᴷ
T
5ᴷ
$M_{R2} = e_2 R_{R2}$
R_{L2}
R
L
R_{R2}
e_2
(b)

FIGURE 8.7

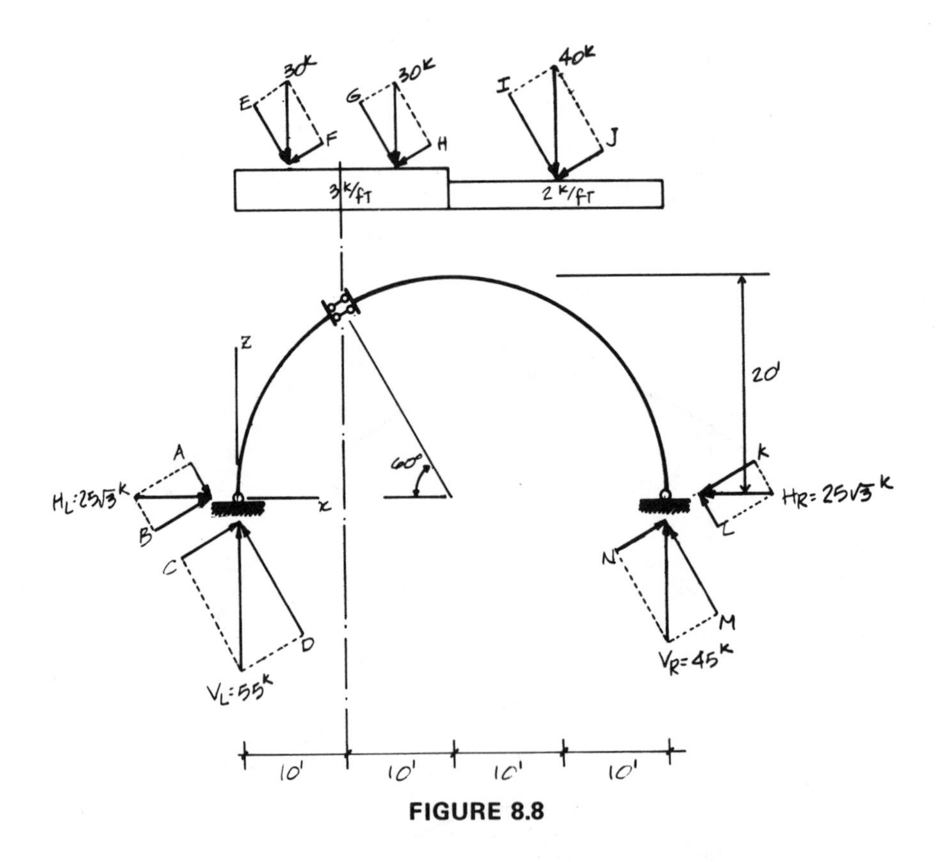

FIGURE 8.8

$$\Sigma M = 0 = -40V_L + 3(20)30 + 2(20)10,$$

which yields

$$V_L = 55 \text{ k}.$$

Second equation: equilibrium of the vertical external forces

$$\Sigma F_y = 0 = V_L - 3(20) - 2(20) + V_R,$$

from which

$$V_R = 60 + 40 - V_L = 45 \text{ k}.$$

Third equation: The shear calculated from the left (or right) end vanishes at the slide connection. Symbolically,

$$A - D + E = 0.$$

Explicitly,

$$H_L \cos 60° - V_L \cos 30° + 3(10) \cos 30° = 0,$$

from which

$$H_L = (V_L - 30)\frac{\cos 30°}{\cos 60°} = 25(3)^{1/2} = 43.3 \text{ k}.$$

Fourth equation: equilibrium of horizontal external forces

$$H_L - H_R = 0.$$

Then,

$$H_R = H_L = 43.3 \text{ k}.$$

If the shear at the slide connection is calculated from the right, the third equation is

$$G + I - M - L = 0;$$

explicitly,

$$3(10) \cos 30° + 2(20) \cos 30° - V_R \cos 30° - H_R \sin 30° = 0,$$

from which

$$H_R = \frac{\cos 30°}{\sin 30°}(30 + 40 - 45) = 25(3)^{1/2} = 43.3 \text{ k}.$$

When the links of the slide are parallel to H_L and H_R, the horizontal reactions do not appear in the third equation, which is therefore useless as stated. A different third equation must state, instead, that the moment at the sliding section calculated from the left equals the moment calculated from the right. In this case, the third and fourth equations are coupled, since they both contain H_L and H_R, and they are solved with any known algebraic procedure.

The graphic solution in Case C is obtained, as in other cases, by using the principle of superposition. When the arch is loadfree from the left hinge to the slide, the line of action of reaction R_{L1} is parallel to the links of the slide drawn from the left hinge (Figure 8.9*a*). Indeed, in this condition only does the shear vanish at the slide and the moment vanish at the left hinge. Reaction R_{R1} of the right hinge is in equilibrium with R_{L1} and the resultant W_R of the

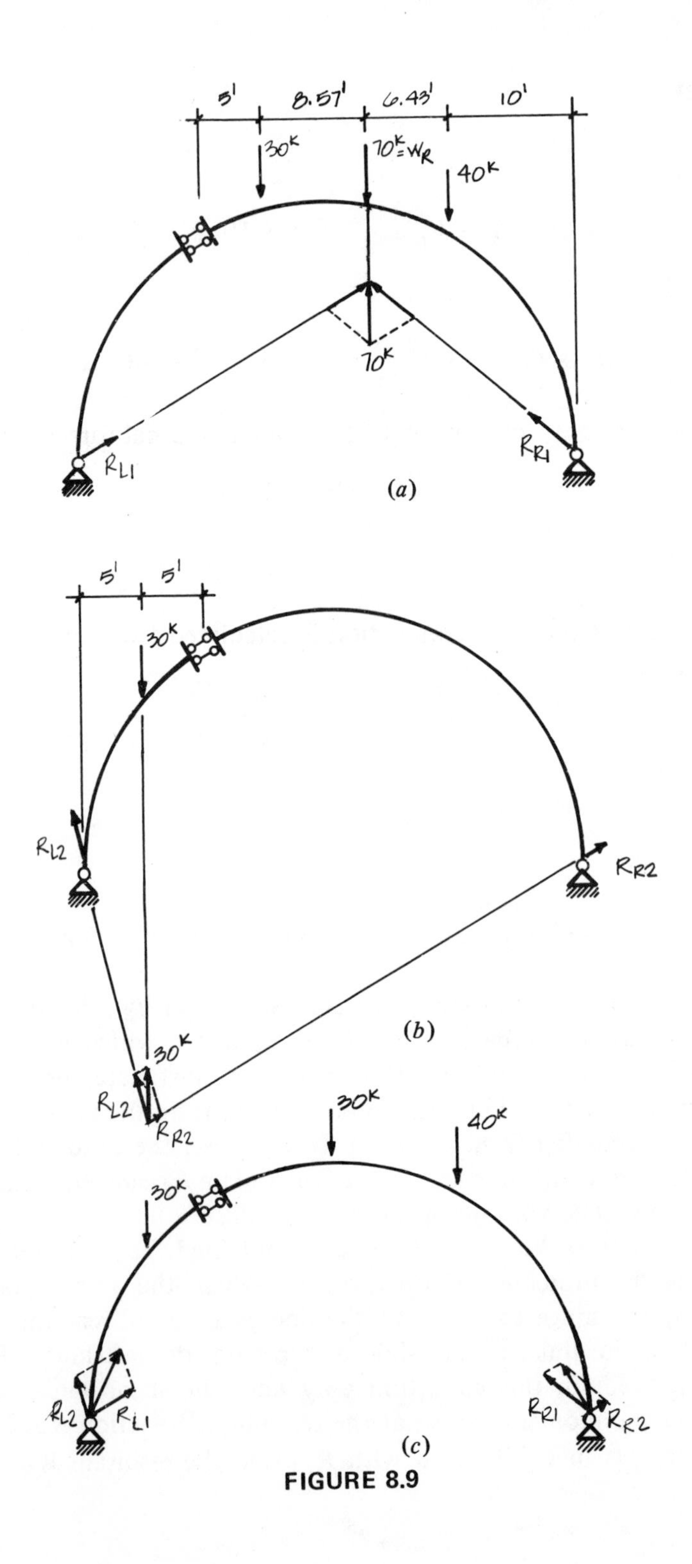
5'
8.57'
6.43'
10'
30K
70K=WR
40K
70K
RL1
RR1
(a)
5'
5'
30K
RL2
RR2
30K
RL2
RR2
(b)
30K
40K
30K
RL2
RL1
RR1
RR2
(c)

FIGURE 8.9

load on the right part of the arch. R_{R1} is therefore concurrent with R_{L1} and W_R. By similar reasoning, we find R_{R2} and R_{L2} (Figure 8.9*b*). R_L and R_R are obtained by the parallelogram rule in Figure 8.9*c*.

Case D

Figure 8.10 shows an arch with sliding connections to its foundations and an intermediate moment release along its axis. The geometries of the axis are those of a circular segment with a 100-ft horizontal span. The left-end section of the arch is inclined 45° on the horizontal plane, and the right-end section is inclined 60° on the same plane. The radius of the circle is given by

$$R \cos 45^\circ + R \cos 60^\circ = 100;$$

thus,

$$R = \frac{100}{1.207} = 82.9 \text{ ft.}$$

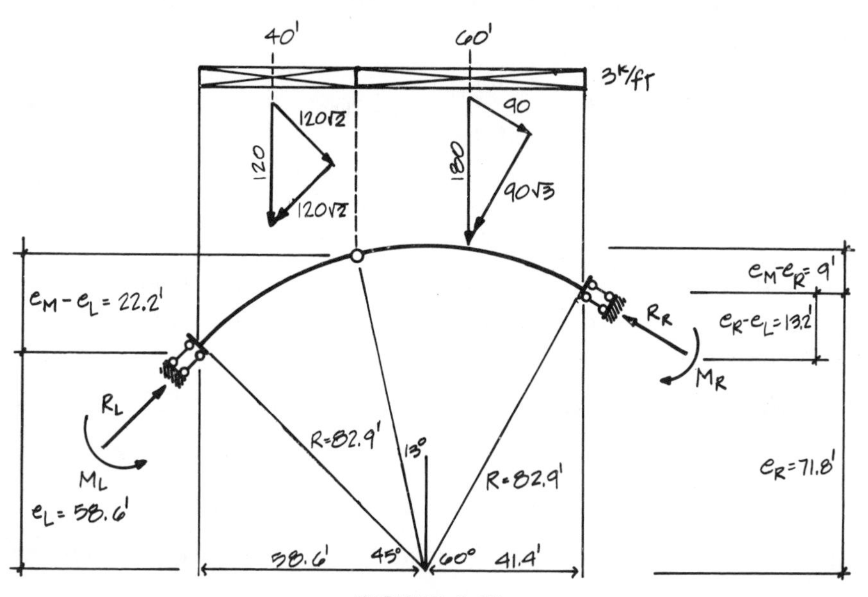

FIGURE 8.10

With respect to the diameter of the circle, elevations of points L, R, and M are, respectively,

$$e_L = R \sin 45° = 82.9(0.707) = 58.6 \text{ ft},$$

$$e_R = R \sin 60° = 82.9(0.866) = 71.8 \text{ ft},$$

$$e_M = [(82.9)^2 - (18.6)^2]^{1/2} = 80.79 \text{ ft}.$$

In this case, reaction R_L has the direction of the links of the left-end slide, and reaction R_R has the direction of the links of the right-end slide. The two reactions are obtained from equations of horizontal and vertical equilibrium for the free body of the arch

$$\Sigma F_x = 0 = R_L \cos 45° - R_R \cos 30°,$$

$$\Sigma F_y = 0 = -300 + R_L \sin 45° + R_R \sin 30°.$$

From the first equation, we obtain

$$R_L = R_R \frac{\cos 30°}{\cos 45°} = 1.225 R_R.$$

Substituting R_L in the second equation yields

$$R_R(\sin 30° + 1.225 \sin 45°) = 300,$$

from which

$$R_R = \frac{300}{0.5 + 0.866} = 219.6 \text{ k}$$

and

$$R_L = 1.225(219.6) = 269 \text{ k}.$$

Moment M_L is obtained by stating that the bending moment calculated from the left end vanishes at the hinge. Assuming that M_L is counterclockwise, the equation is

$$M_L + (e_M - e_L) R_L \cos 45° - 40 R_L \sin 45° + 20(120) = 0,$$

from which

$$M_L = 0.707(40 - 22.2)\, 269 - 2400 = 985 \text{ k-ft}.$$

Similarly, moment M_R is obtained by stating that the bending

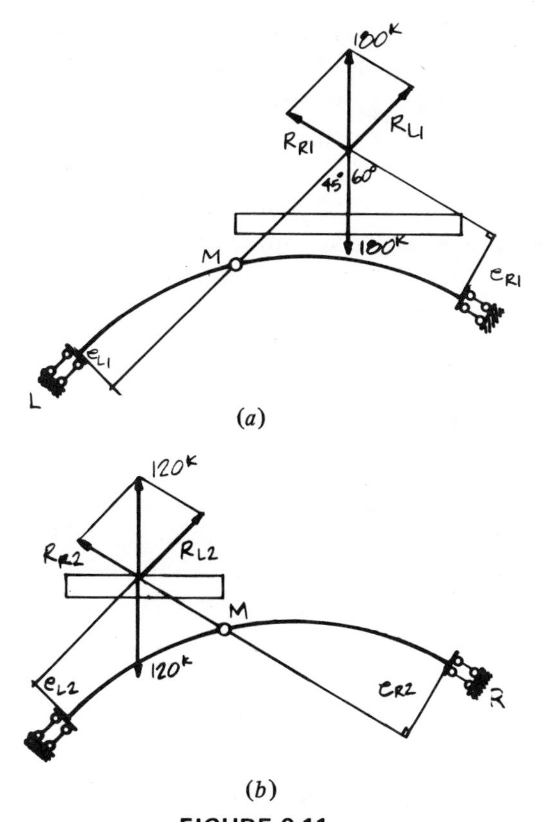

FIGURE 8.11

moment calculated from the right vanishes at the hinge. M_R is assumed to be clockwise. Then, the equation is

$$-M_R - (e_M - e_R)\,R_R \cos 30° + 60R_R \sin 30° - 30(180) = 0,$$

from which

$$M_R = -9(219.6)\,0.866 + 60(219.6)\,0.5 - 5400 = -523.6 \text{ k-ft}.$$

The negative sign in the result suggests that M_R is counterclockwise rather than clockwise, as assumed.

In the graphic solution of this case, the load on part LM of the arch is initially removed (Figure 8.11*a*). Then, the interaction at M has the direction of the links of the slide at L. It is, indeed, a necessary condition that the shear vanish at L and the moment vanish at M. R_{L1} and R_{R1} are in equilibrium with the 180-k resultant of the load on part MR of the arch. Therefore, R_{R1} is concurrent with R_{L1} and the 180-k force, and it has the direction of the links of slide R. Reactive moments M_{L1} and M_{R1} are the products of R_{L1} and R_{R1} with their eccentricities from points L and M, respectively. Figure 8.11*b* shows the analogous graphic evaluation of R_{L2}, R_{R2}, M_{L2}, and M_{R2}. The combination of partial reactions yields the reaction to the complete systems of loads according to the principle of superposition.

Case E

An arch shaped as a circular segment has the geometries, constraints, and loads shown in Figure 8.12. Reaction R_R of the right hinge has component H_R perpendicular to the links of the slide at L and component V_R parallel to them. In evaluating H_R, it is necessary to state that the shear vanishes at point L

$$H_R - 300 \cos 60° = 0,$$

thus,

$$H_R = 150 \text{ k}.$$

V_R is found by stating that the shear force calculated on the right side of point M vanishes at M.

$$H_R \cos 60° + V_R \cos 30° - 150 = 0,$$

from which

$$V_R = \frac{2}{(3)^{1/2}}(150 - 75) = 86.6 \text{ k},$$

$$R_R = (H_R^2 + V_R^2)^{1/2} = (150^2 + 86.6^2)^{1/2} = 173.2 \text{ k}.$$

An equation of equilibrium in the direction of R_L yields its value

$$R_L - 300 \cos 30° + V_R = 0,$$

thus,

$$R_L = 259.8 - 86.6 = 173.2 \text{ k}.$$

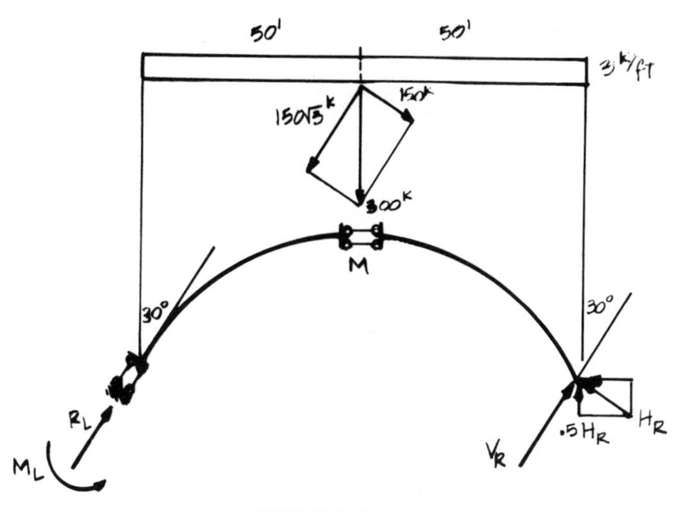

FIGURE 8.12

Finally,

$$M_L = -100(H_R \cos 60^\circ + V_R \cos 30^\circ) + 50(300)$$
$$= 15000 - 100(50 + 75) = 2500 \text{ k-ft.}$$

In the graphic evaluation of the reactions, reaction R_{R1} of the hinge (Figure 8.13*a*) must be parallel to the links of the slide at M in order to eliminate the shear force on section M. Since R_{L1} is in equilibrium with R_{R1} and the 150-k load on segment LM of the arch, the line of action of R_{L1} is concurrent with those of R_{R1} and the 150-k load.

When segment MR of the arch is loaded and segment LM is loadfree (Figure 8.13*b*), reaction R_{L2} should be parallel to the links of the slides at L and M in order to eliminate the shear at both points. This is possible only if either sections L and M are parallel, in which case the arch is unstable or R_{L2} equals zero. Therefore, when segment LM is loadfree, the slide at L reacts only with moment M_{L2}, which is transferred unchanged along LM to slide M. Reaction R_{R2} of the hinge must be equal, opposite, and parallel to the resultant of the load on segment MR of the arch, thus eliminating any action on segment LM except the moment of the couple.

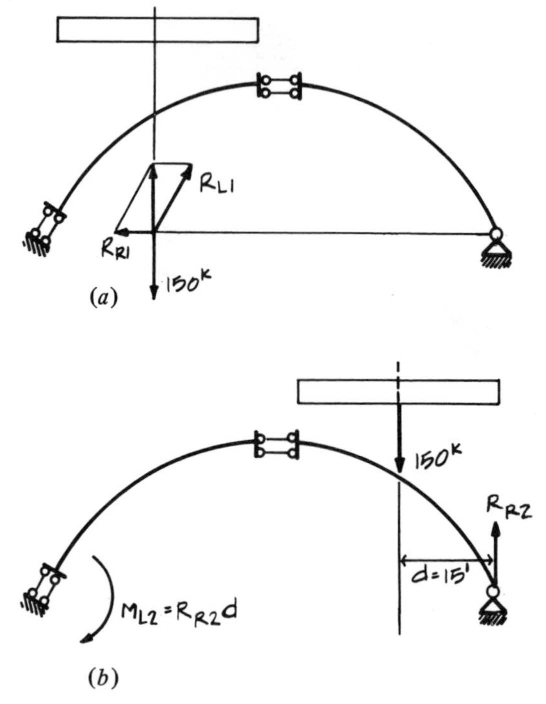

FIGURE 8.13

Case F

This case is readily understood by comparing it with Case D. To discuss it, therefore, we use the arch in Figure 8.11 with the same geometries, load, and external constraints but with the hinge replaced by a third slide (Figure 8.14). Reactions R_L and R_R are obtained from the two equations of horizontal and vertical equilibrium for the free body of the arch.

Since the geometries, loads, and external constraints shown in Figure 8.14 are identical to those shown in Figure 8.11 equations of force equilibrium are necessarily identical in the two cases and yield identical results

$$R_L = 269 \text{ k},$$

$$R_R = 219.6 \text{ k}$$

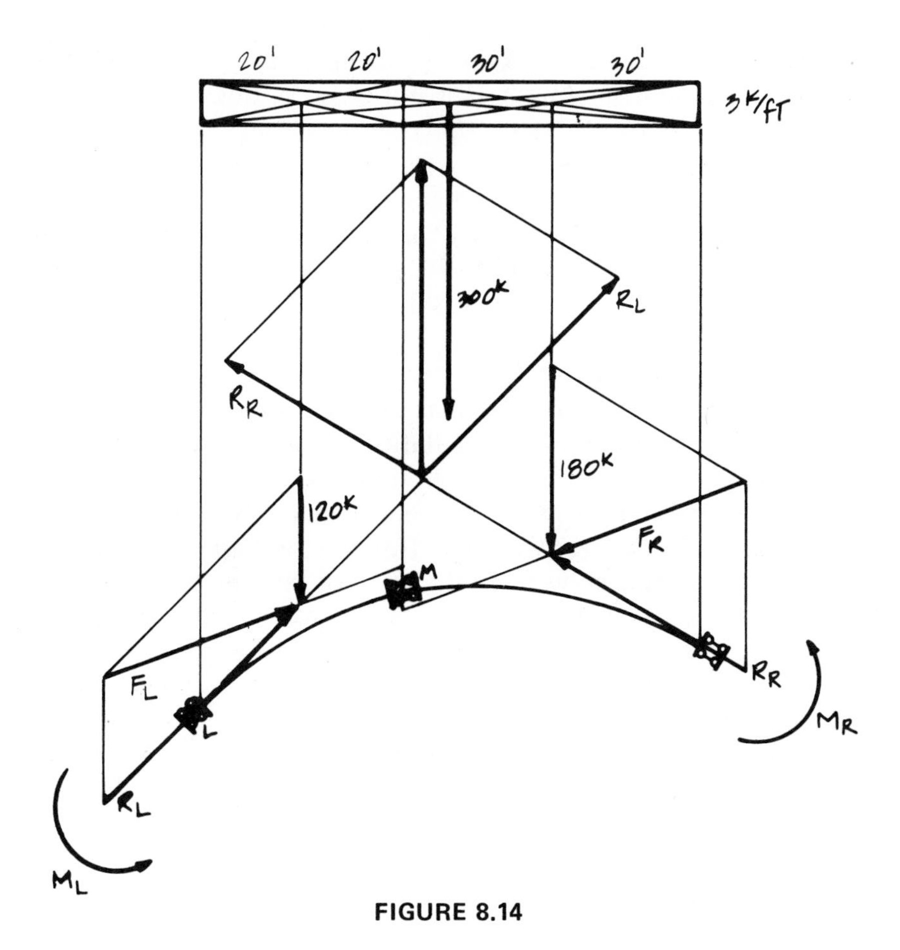

FIGURE 8.14

In Case D, reactive moment M_L is the moment necessary to shift resultant F_L of reaction R_L with the 120-k load on LM to a lower line of action containing point M (Figure 8.14), thus eliminating bending at the hinged section. Indeed, in Case D, the counterclockwise moment $M_L = 985$ k-ft was numerically obtained by stating in equation form that bending should vanish at point M. The counterclockwise moment $M_R = 523.6$ k-ft is obtained with analogous graphic or numerical operations.

In the new case F, the sum of M_L and M_R must, again, be equal to

$$M_L + M_R = 985 + 523.6 = 1508.6 \text{ k-ft}$$

in order to balance the clockwise moment of the couple of 300-k loading and equilibrant forces. It is not possible, however, to obtain individual values for M_L and M_R from conditions of equilibrium that are graphically or numerically stated. Indeed, moment M_L must shift the line of action of F_L, and moment M_R must shift the line of action of F_R, so that F_L and F_R are colinear on a unique line of action parallel to the links of the slide at M. It is not possible, however, to determine the individual shift of F_L and F_R from equilibrium conditions; rather, $M_L + M_R$ must eliminate the total arm of the couple of forces F_L and F_R.

We must, therefore, conclude that from the viewpoint of bending moments, the arch in Case F is a statically indeterminate structure. The string polygon or reversed cable shape, as in Case D, has its first side colinear with R_L, its second side colinear with F_L and F_R, and its third side colinear with R_R. However, in Case D, the correct position of the string polygon on the plane of the structure is defined by the condition that its second side must cross the axis of the arch at the hinged section M, while in Case F, an analogous condition is lacking. Shear and Axial forces are identical in Cases D and F, since external forces (loads and reactions) are identical. Reactive moments M_L and M_R, and thus internal bending moments, are different in the two cases, and in Case F, they can not be determined by statics.

PROBLEMS

8.1. Find the reactions of the three-hinges arch in Figure 8.15.

Solution.

$$V_L = 48.4 \text{ k},$$

$$H_L = 27.4 \text{ k},$$

$$V_R = 11.6 \text{ k},$$

$$H_R = 7.4 \text{ k}.$$

8.2. Using the graphic procedure with scales S_F and S_D of the reader's own choice, find reactions R_L and R_R of the three-

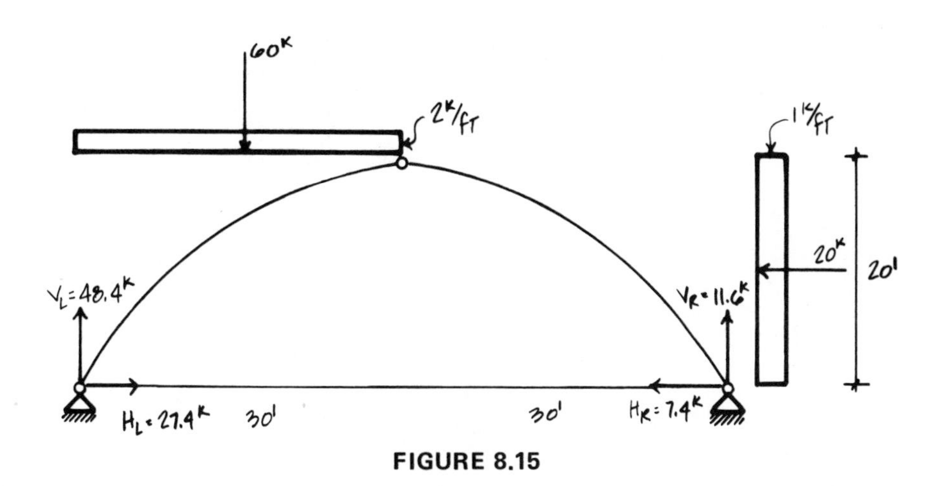

FIGURE 8.15

hinges arch in Figure 8.15. Check the vertical and horizontal components of R_L and R_R with the answers in Problem 8.1.

8.3. Find axial force A and shear force S at the hinged crown of the arch (Figure 8.15).

Solution. At the specified section, the axial force is horizontal. Summing all horizontal forces from the left end to midspan, we obtain

$$A = H_L = 27.4 \text{ k.}$$

Summing all horizontal forces from the right end again, we obtain

$$A = H_R + 1(20) = 27.4 \text{ k.}$$

The shear force is vertical at the specified section. Summing all vertical forces from the left end, we obtain

$$S = V_L - 2(30) = 11.6 \text{ k.} \quad \text{(down)}$$

Summing all vertical forces from the right end again, we obtain

$$S = V_R = 11.6 \text{ k.} \quad \text{(up)}$$

8.4. Graphically obtain the resultant of external forces (loads and reactions in Problem 8.2) on one side of the crown hinge. Decompose this resultant into axial force A and shear force

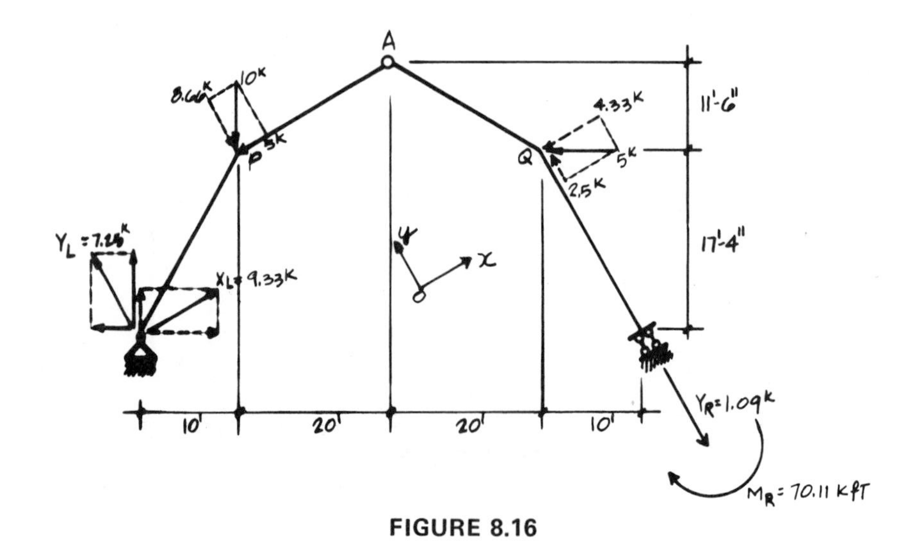

FIGURE 8.16

S at the crown hinge. Check the graphic results with the answers in Problem 8.3. Use scales S_F and S_D of your own choice (arch in Figure 8.15).

8.5. Find the reactions of the constraints of the arch in Figure 8.16.

Solution. The y-axis joining the pivot of the left end to that of the right end is the parallel to the links of the right end drawn from the left-end hinge. The right-end section can, indeed, translate only in the x direction. This translation can be called a rotation around a center infinitely far in the direction of the links. In the chosen xy frame of orthogonal axes, reactions of the hinge are X_L and Y_L. The slide reacts with a force Y_R and a moment M_R. X_L is obtained from the equation

$$\Sigma F_x = 0 = X_L - 5 - 4.33,$$

from which

$$X_L = 9.33 \text{ k}.$$

Using these notations for the vertical and horizontal components of X_L and Y_L,

$$X_{Lv} = X_L \sin 30^\circ = 0.5\, X_L,$$

$$X_{Lh} = X_L \cos 30° = 0.866\, X_L,$$

$$Y_{Lv} = Y_L \cos 30° = 0.866\, Y_L,$$

$$Y_{Lh} = Y_L \sin 30° = 0.5\, Y_L,$$

we state in equation form that the moment of all forces from the left end to the crown hinge vanishes at the crown hinge

$$\Sigma M = 0 = -30(X_{Lv} + Y_{Lv}) + 28.82(X_{Lh} - Y_{Lh}) + 20(10).$$

Replacing symbols with numbers,

$$0 = -30(0.5)9.33 - 30(0.866)\, Y_L + 28.82(0.866)9.33$$

$$- 28.82(0.5)\, Y_L + 200,$$

from which

$$Y_L(25.98 + 14.42) = -139.95 + 232.86 + 200.$$

Solving for Y_L

$$Y_L = \frac{292.91}{40.4} = 7.25 \text{ k}.$$

The equation of equilibrium in the y direction yields

$$Y_R = 7.25 - 8.66 + 2.5 = 1.09 \text{ k}.$$

Using these notations for the vertical and horizontal components of Y_R,

$$Y_{Rv} = Y_R \cos 30° = 0.866\, Y_R = 0.944 \text{ k},$$

$$Y_{Rh} = Y_R \sin 30° = 0.5 \quad Y_R = 0.545 \text{ k},$$

we state in equation form that the sum of all moments from the right end to the crown hinge around the crown hinge vanishes:

$$\Sigma M = 0 = -M_R - 30Y_{Rv} + 28.32\, Y_{Rh} - 5(11.5).$$

Replacing symbols with numbers,

$$0 = -M_R - 30(0.944) + 28.82(0.545) - 57.5,$$

from which

$$M_R = -28.32 + 15.71 - 57.50 = -\ 70.11 \text{ k}.$$

The negative result requires reversing the sign initially assumed for M_R.

8.6. Using the graphic procedure and scales S_F and S_D of your own choice, find the reactions of the constraints of the arch in Figure 8.16. Check results with the answers in Problem 8.5.

8.7. Find axial force A, shear force S, and bending moment M at section P of the arch in Figure 8.16.

Solution. Starting from the left end,

$$A_P = X_L - 5 = 9.33 - 5 = 4.33 \text{ k},$$

$$S_P = Y_L - 8.66 = 7.25 - 8.66 = -1.41 \text{ k},$$

$$M_P = -10(Y_{Lv} + X_{Lv}) - 17.32(Y_{Lh} - X_{Lh}).$$

Replacing symbols with their numerical values,

$$M = -10(6.28 + 4.67) - 17.32(3.63 - 8.08) = -32.43 \text{ k-ft}.$$

Starting from the right end,

$$A_P = -4.33 \text{ k},$$

$$S_P = -Y_R + 2.5 = -1.09 + 2.5 = 1.41 \text{ k}.$$

With the corrected sign of M_R,

$$M_P = M_R - 50Y_{Rv} + 17.32Y_{Rh}.$$

Replacing symbols with numbers,

$$M = 70.1 - 50(0.944) + 17.32(0.545) = 32.35 \text{ k-ft}.$$

8.8. Graphically perform the tasks in Problem 8.7. Use the same scales S_F and S_D as in Problem 8.6. Compare the graphic results with the numerical answers in Problem 8.7.

8.9. Find axial force A, shear force S, and bending moment M at section Q of the arch in Figure 8.16.

8.10. Graphically perform the tasks in Problem 8.9. Use the same scales S_F and S_D as in Problem 8.6. Compare the graphic results with the numerical answers in Problem 8.9.

8.11. Place the 10-k vertical load on the arch in Figure 8.16 at

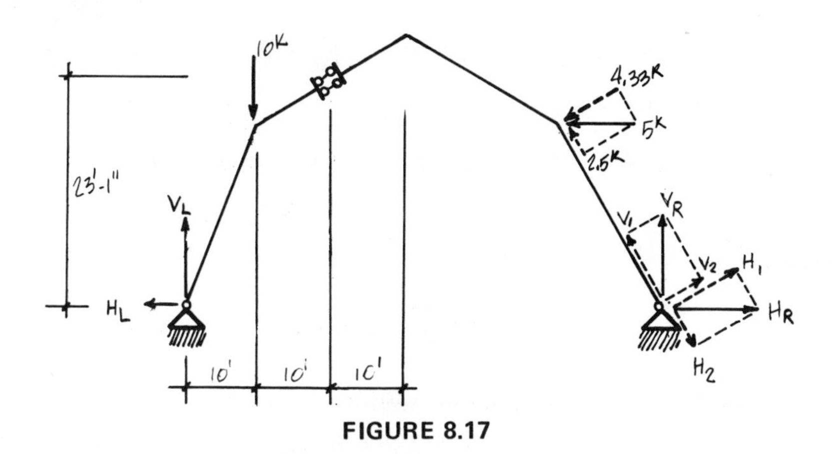

FIGURE 8.17

point Q and the 5-k horizontal load at point P. Find Y_L, X_L, Y_R, M_R, A_P, S_P, M_P, A_Q, S_Q, and M_Q.

8.12. Using scales S_F and S_D as in Problem 8.6 or scales of your own choice, graphically perform the tasks in Problem 8.11.

8.13. Numerically find the reactions of the constraints of the truss-arch in Figure 7.20. The solution is shown in Figure 7.20.

8.14. Numerically find the reactions of the constraints of the truss-arch in Figure 7.24*a*. The solution is shown in Figure 7.24*a*.

8.15. Find reactions V_L, H_L, V_R, and H_R of the arch in Figure 8.17.

Solution. The equation of moment equilibrium around the left hinge states

$$\Sigma M = 0 = -10(10) + 5(17.32) + 60V_R,$$

from which

$$V_R = \frac{100 - 86.6}{60} = 0.223 \text{ k}.$$

From the equation of vertical equilibrium, we obtain

$$V_L = 10 - 0.223 = 9.777 \text{ k}.$$

An equation of force equilibrium in a direction orthogonal to the links of the slide can be written with the components of V_R and H_R

$$V_1 = V_R \cos 30^\circ = 0.866\,V_R = 0.193 \text{ k},$$

$$H_2 = H_R \cos 60^\circ = 0.5H_R$$

and with the components of the loads shown in Figure 8.17

$$\Sigma F = 0 = 2.5 + 0.193 - 0.5H_R,$$

from which

$$H_R = \frac{2.693}{0.5} = 5.386 \text{ k}.$$

From the equation of horizontal equilibrium, we obtain

$$H_L = -5 + 5.386 = 0.386 \text{ k}.$$

8.16. Following the graphic procedure in Figure 8.9, find the reaction of the arch in Figure 8.17. Use scales S_F and S_D of your own choice.

8.17. Using external forces or the components shown in Figure 8.17 on the right side of the slide, find axial force A and bending moment M in the slide.

Solution.

$$M = 40V_R + 23.17H_R - 5(5.75) = 40(0.223) + 23.17(5.386) - 28.75$$
$$= 105 \text{ k-ft}$$

$$A = H_1 + V_2 - 4.33 = 0.866H_R + 0.5V_R - 4.33 = 0.445 \text{ k}.$$

8.18. Check the solution to Problem 8.17 by finding values for M and A with external forces on the left side of the slide.

8.19. Find reactions A_L, A_R, M_L, and M_R of the constraints of the arch in Figure 8.18.

Solution. With the components of external forces perpendicular to A_L, the equation of equilibrium in the direction orthogonal to A_L states

$$A_2 + 4.33 - 5 = 0,$$

from which

$$A_2 = 0.67 \text{ k}.$$

Thus,

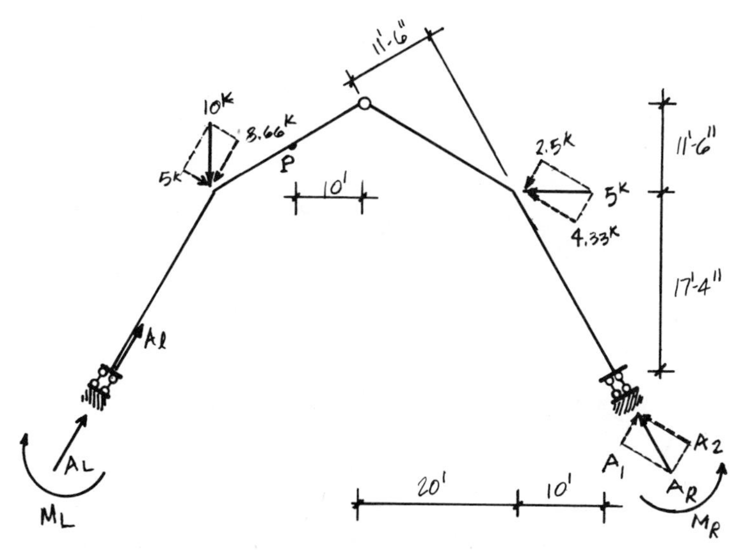

FIGURE 8.18

$$A_R = \frac{A_2}{\cos 30^\circ} = 0.77 \text{ k}.$$

With external forces and moment on the right side of the crown hinge, we can state that bending vanishes at the crown hinge

$$M_R + 11.5A_R - 5(11.5) = 0,$$

from which

$$M_R = 57.5 - 8.86 = 48.6 \text{ k-ft}.$$

The equation of equilibrium in the direction of A_1 states

$$A_L - 8.66 - 2.5 + A_1 = 0,$$

from which

$$A_L = 8.66 + 2.5 - 0.385 = 10.775 \text{ k}.$$

With external forces and moment on the left side of the crown hinge, we can state that bending vanishes at the crown hinge

$$-M_L - 11.5A_L + 20(10) = 0,$$

from which

$$M_L = 200 - 123.9 = 76.1 \text{ k-ft}.$$

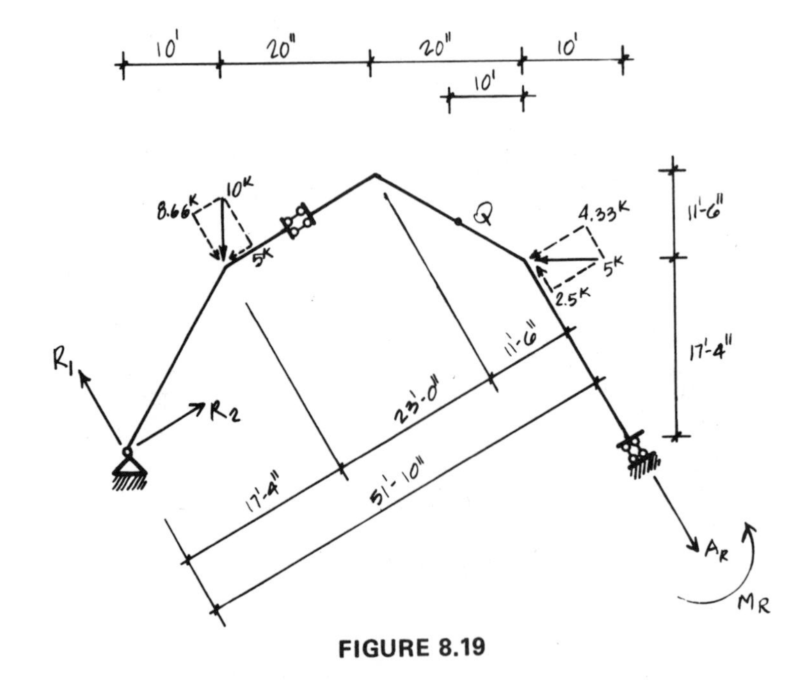

FIGURE 8.19

8.20. Perform the tasks in Problem 8.19 using the graphic procedure in Figure 8.11.

8.21. With external forces and moments on the left of point P, find internal forces A_P and S_P and the internal moment M_P. Check the results by solving the same problem but using external forces and moments on the right of P (Figure 8.18).

8.22. Find reactions R_L, A_R, and M_R of the constraints of the arch in Figure 8.19.

Solution. With external forces on the left of the internal slide, we state that the shear vanishes at the internal slide

$$R_1 - 8.66 = 0,$$

from which

$$R_1 = 8.66 \text{ k}.$$

The shear vanishes at the terminal slide. Thus,

$$R_2 - 5 - 4.33 = 0,$$

from which

$$R_2 = 9.33 \text{ k}$$

and

$$R_L = (R_1^2 + R_2^2)^{1/2} = (8.66^2 + 9.33^2)^{1/2} = 12.73 \text{ k}.$$

The equation of equilibrium in the direction R_1 states

$$R_1 - 8.66 + 2.5 + A_R = 0,$$

from which

$$A_R = 2.5 \text{ k}.$$

The equation of moment equilibrium around the hinge states

$$\Sigma M = 0 = -10(10) + 5(17.32) - 51.82A_R + M_R,$$

from which

$$M_R = 100 - 86.6 + 129.55 = 142.95 \text{ k-ft}.$$

8.23. Using the graphic procedure in Figure 8.13, perform the tasks in Problem 8.22.

8.24. With external forces and moments on the left of point Q, find internal forces A_Q, S_Q, and internal moment M_Q. Check the results by solving the same problem using external forces and moments on the right of point Q.

View of the cross section of the planks of a concrete floor. Photograph courtesy of Flexicore Co., Inc., Dayton, Ohio.

NINE

GEOMETRIC PROPERTIES OF CROSS SECTIONS

Evaluating the internal stresses in a structural element requires previous knowledge of the internal forces as well as a knowledge of the properties of the cross section that condition effects of the internal forces. For example, the numerical value of the area A of the cross section of a cable determines the value of the stress P/A that the axial force P produces on a cable section.

Similarly, the magnitude of the shear, bending, twisting, and buckling stresses produced on a section by the local internal forces is determined by other cross-sectional properties. Specifically, the value of the bending stress is conditioned by a cross-sectional property called the *moment of inertia of the section around the centroidal axis of revolution of the bending moment.* This property will be referred to as the section's moment of inertia. Evaluating the bending stress (a Step-5 task) must, therefore, be preceded by recognizing the centroidal axis and evaluating the moment of inertia around this axis (Step-4 tasks).

The magnitude of the shear stress is conditioned by a cross-sectional property that is the distance between the resultant of compressive stresses and the resultant of tensile stresses produced by bending. This distance is called the *lever arm of the resisting moment.* The twisting stress on circular sections is determined by a geometric property called the *polar moment of inertia.* The buckling stress is determined by a geometric property of the cross section called the *minimum radius of gyration.* Defining procedures and formulas for evaluating these properties is the objective of this chapter.

9.1. THE AREA OF A CROSS SECTION

The cross section of a structural element has, in most cases, a simple, polygonal shape that permits easy evaluation of the area or (in the case of prefabricated elements) complex outlines associated with the most efficient distribution of structural material on the cross section. In the latter case, manuals issued by manufacturers of the prefabricated elements give the value of the cross-sectional area along with other relevant properties. In general, however, an area can be easily calculated when the area's infinitesimal element is a recognizable function of the reference coordinates.

For example, an infinitesimal element of the triangular section in Figure 9.1*a* has width b, which is the linear function $(B/H)\,z$ of the z-coordinate of the element. The area A can, therefore, be calculated as the sum of the infinite number of infinitesimal elements $dA = bdz$. That sum is the integral

$$A = \int_{z=0}^{z=H} dA = \int_0^H b\,dz = \frac{B}{H}\int_0^H z\,dz = \frac{BH}{2}.$$

Similarly, the infinitesimal element of the circular section in Figure 9.1*b* has length $2\pi r$, which is a function of the polar coordinate r. The area of the circle is then

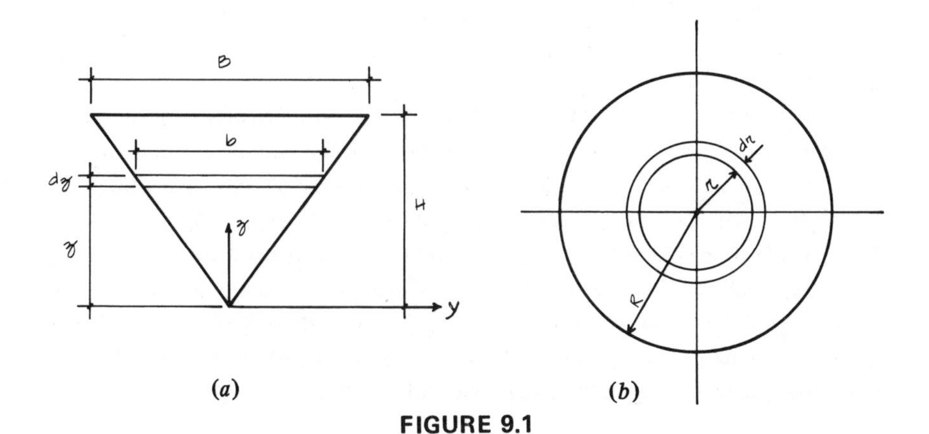

(*a*) (*b*)

FIGURE 9.1

$$\int_{r=0}^{r=R} dA = 2\pi \int_0^R r\,dr = \pi R^2.$$

When the area does not have a ruled outline, it must be drawn in scale, preferably on grid paper, and it is then measured by using either an instrument called a planimeter or by counting the number of grid subdivisions enclosed in the area's outline.

9.2. AREA MOMENTS, THE CENTER OF A CROSS SECTION

A centroidal axis of the cross section of a structural element is a line through the center of the area. The area center is found by an analytic procedure analogous to experimentally finding the center of mass C of a thin disk of structural material. After attaching a string to a point on the disk and hanging the disk by the string, a pendular motion takes the disk to a position where its weight W is colinear with the string (Figures 9.2a and b). Then, a segment colinear with the string and with the weight is drawn on the disk; this segment contains the center of mass C. Attaching the string to a different point on the disk and repeating the experiment, we draw a second line that also contains the center of mass C. The intersection of the two lines is the point in question.

As in the experimental procedure, we must draw two centroidal

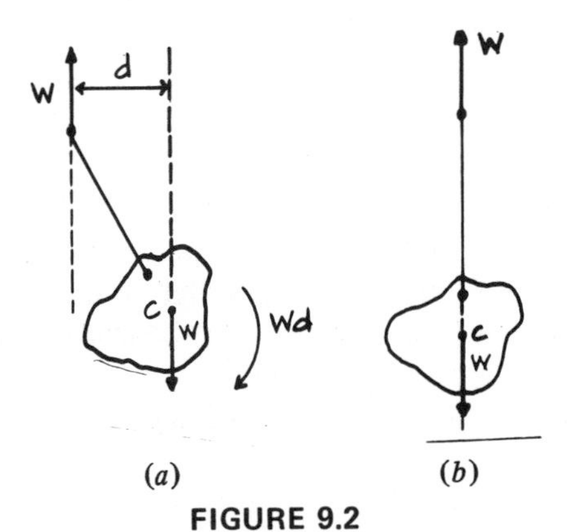

FIGURE 9.2

axes on the cross section in order to locate the center of the section at their intersection. Since an axis of geometric symmetry contains the center of the section, it is only necessary to find one other centroidal axis when such an axis exists. Its intersection with the axis of symmetry is the center of the cross section. If two axes of symmetry exist, they intersect one another at the center of the section.

Centroidal axes other than axes of symmetry are found by the same technique that gives the line of action of the resultant (total weight) of a system of parallel forces (partial weights). A cross-sectional area could, indeed, be considered the leftover of the mass of a slice of a structural element as the thickness of the slice, and therefore its volume, vanishes. Then, the center of mass is reduced to the center of the area.

This technique requires dividing the cross-sectional area into partial areas having two axes of symmetry and, therefore, individual centers known by inspection. The partial areas are treated like forces parallel to the desired centroidal axis of the section. The magnitude of these forces equals the value of the partial areas.

Cavities in the cross section are considered as negative forces. An example is the area of the inner circle of the section of a pipe. The total area is the resultant of the partial areas. It must, therefore, satisfy the same equivalence requisites for a resultant force.

The resultant of a system of forces is a unique force equivalent to the given system, which means it must produce the same state of rest or motion that the system produces on the structure to which it is applied. To be equivalent in translation, the resultant must be the vectorial sum of the components. To be equivalent in rotation, the resultant must have a moment equal to the sum of the moments of the components around any pivot point.

Applying the first condition of equivalence to a system of partial areas, we obtain the total area by summing the parts. Since all the areas are represented by vectors parallel to the desired centroidal axis, the vectorial sum coincides with the algebraic sum of the parts. The centroidal axis is found by the second condition of equivalence as the line of action of the resultant, which contains its point of application.

The T section in Figure 9.3a, made by welding two steel plates,

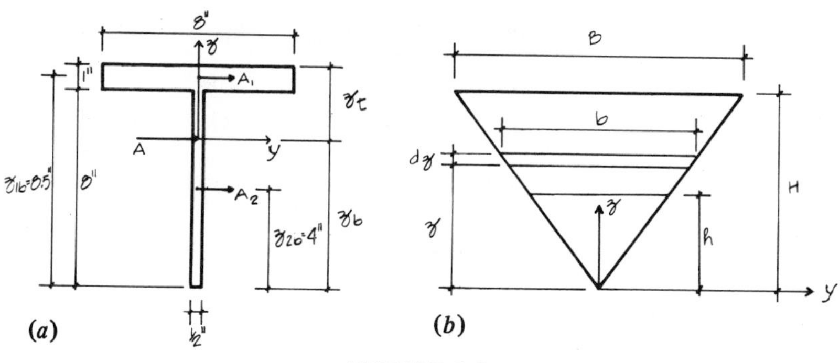

FIGURE 9.3

demonstrates this technique. The flange plate and the web plate both have two axes of symmetry, and they therefore have known individual centroids. The partial areas are $A_1 = 8$ in.^{2} and $A_2 =$ 4 in.^{2} The section has a vertical axis of symmetry that contains the center C of the area. The horizontal centroidal axis is not known and must be found. Two horizontal area vectors A_1 and A_2 are applied to the centers of the partial areas: A_1 to the center of the flange; A_2 to the center of the web. The total area vector A is the resultant of the partial area vectors A_1 and A_2. By the first condition of equivalence,

$$A = A_1 + A_2 = 8 + 4 = 12 \text{ in.}^2$$

By the second condition of equivalence, the moment of A around any pivot point—for instance, around a bottom fiber of the section—equals the sum of the moments of A_1 and A_2 around the same point.

$$(A_1 + A_2)\, z_b = A_1 z_{1b} + A_2 z_{2b} = 8(8.5) + 4(4) = 84 \text{ in.}^3$$

Dividing by $A_1 + A_2$, we get

$$z_b = \frac{A_1 z_{1b} + A_2 z_{2b}}{A_1 + A_2} = \frac{\sum_{i=1}^{2} A_i z_{ib}}{\sum_{i=1}^{2} A_i} = \frac{84}{12} = 7.0 \text{ in.}$$

For a large number of partial areas, it may be convenient to organize operations and results in a table like the following one.

i	A_i in.^{2}	z_{ib} in.	$A_i z_{ib}$ in.^{3}
1	8	8.5	68
2	4	4	16
$\sum_{i=1}^{2}$	12		84

The summation

$$Q_b = \sum_{i=1}^{n} A_i z_{ib}$$

is called the static moment of the area A or the first-order area moment around the bottom fibers of the section. The expression static moment indicates that an area moment does not produce rotation like the moment of a force.

A cross-sectional area may not be made of finite parts with recognizable magnitude and center (Figure 9.3*b*). If, however, the section has an outline that allows us to express the elemental area as a known function of its elevation, the coordinate z, then the centroidal axis can be found by applying the same technique used for finite partial areas to the elemental areas.

The triangular section in Figure 9.3*b* is used for demonstration. The elemental area dA has width b, which is the linear function $(B/H)\,z$ of the z-coordinate. The magnitude of the area dA is, then, $b\,dz = (B/H)\,z\,dz$. It is not necessary to know the center of the elemental area. Its distance from the reference axis y is, in general, given by $z + f\,dz$, where $f\,dz$ is a fraction of the depth dz of the elemental area. Since z is a finite distance and $f\,dz$ is infinitesimal, the latter is negligible by comparison with the former.

The static moment of the partial area is, therefore,

$$dQ_b = z\,dA = \frac{B}{H} z^2\,dz.$$

The sum of the infinite number of infinitesimal moments dQ is equated to the moment of the total area, which is

$$z_b A = \int_{z=0}^{z=H} dQ_b.$$

Writing this equation in the form

$$z_b \int_0^H dA = \int_0^H z\, dA,$$

we obtain in our case

$$z_b = \frac{\int_0^H z\, dA}{\int_0^H dA} = \frac{(B/H)\int_0^H z^2\, dz}{(B/H)\int_0^H z\, dz} = \frac{H^3/3}{H^2/2} = \frac{2}{3}H.$$

The z-coordinate of the center of a trapezoid with its large base B at elevation H and its small base at elevation h is obtained similarly by

$$z = \frac{(B/H)\int_h^H z^2\, dz}{(B/H)\int_h^H z\, dz} = \frac{(1/3)\,(H^3 - h^3)}{(1/2)\,(H^2 - h^2)}.$$

After finding the distance $2H/3$ from the center to the vertex of the triangle, we can also find the z-coordinate of the center of the trapezoid by operating on two triangular partial areas, which are

$$A_1 = \frac{BH}{2},$$

$$A_2 = -\frac{h}{2}\left(\frac{B}{H}h\right).$$

In this case, the second requisite of static equivalence is

$$z\left(\frac{BH}{2} - \frac{hB}{2H}h\right) = \frac{2H}{3}\left(\frac{BH}{2}\right) - \frac{2h}{3}\left(\frac{hB}{2H}h\right),$$

from which

$$z\frac{B}{2H}(H^2 - h^2) = \frac{B}{3H}(H^3 - h^3).$$

Then, again,

$$z = \frac{2(H^3 - h^3)}{3(H^2 - h^2)}.$$

For example, when $H = 18$ in. and $h = 6$ in.,

$$z = 13 \text{ in.}$$

9.3. SECOND-ORDER AREA MOMENTS, THE CENTER OF AN AREA MOMENT

The moment of the static moment is called the second-order moment. When the reference axis of the static moment and the reference axis of the second-order moment are chosen to be the same, the second-order moment is called the moment of inertia. The static moment is given by the product of the area A and the distance Z_A from the reference axis y to the center C_A of the area (Figure 9.4a).

$$Q = AZ_A.$$

If the area and its center are not known otherwise, Q is calculated by $\int z\, dA$, and A is calculated by $\int dA$; then,

$$Z_A = \frac{Q}{A} = \frac{\int z\, dA}{\int dA}.$$

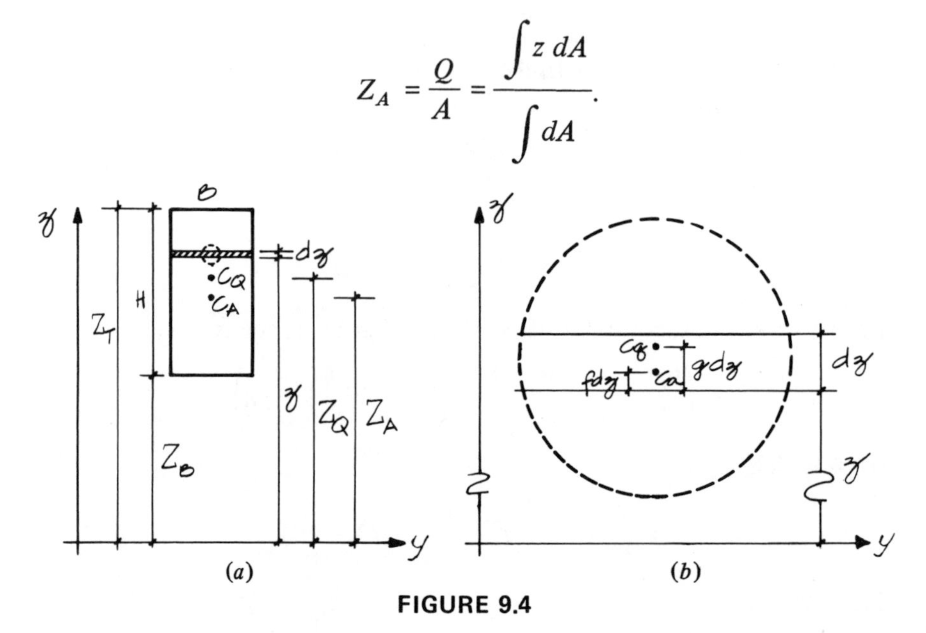

FIGURE 9.4

Following the reasoning in the examples in Section 9.2, we define the moment of inertia as the product of the static moment Q and the distance Z_Q from the reference axis to the center C_Q of the static moment

$$I = QZ_Q = AZ_A Z_Q.$$

The center of the static moment Q is seldom known, since, among other things, it varies as Q varies with the choice of the reference axis. It is therefore necessary to calculate the moment of inertia I as the sum $\int dI$ of the infinite number of infinitesimal moments of inertia dI.

In this case, it is not necessary to know the exact coordinate of the center of the static moment dQ, since this coordinate equals that of the base of the elemental area dA plus a fraction $g\, dz$ of its negligible depth dz (Figure 9.4*b*). Therefore, we can give the elemental moment of inertia a simpler definition

$$dI = z\, dQ = z(z\, dA) = z^2\, dA.$$

Using this definition, the moment of inertia of the entire area is given by

$$I = \int_A dI = \int_A z^2\, dA.$$

The center of the static moment Q is found by

$$Z_Q = \frac{I}{Q} = \frac{\int_A z^2\, dA}{\int_A z\, dA}.$$

Taking the rectangle in Figure 9.4*a* as an example, we have

$$Q = \int_{Z_B}^{Z_T} z\, dA = B \int_{Z_B}^{Z_T} z\, dz = \frac{B}{2}(Z_T^2 - Z_B^2),$$

$$I = \int_{Z_B}^{Z_T} z^2\, dA = B \int_{Z_B}^{Z_T} z^2\, dz = \frac{B}{3}(Z_T^3 - Z_B^3),$$

$$Z_Q = \frac{I}{Q} = \frac{2}{3}\left(\frac{Z_T^3 - Z_B^3}{Z_T^2 - Z_B^2}\right) .$$

With the values $B = 6$ in., $H = 12$ in., and various values for Z_B, the values of Z_A, Q, I, and Z_Q are calculated to show how these properties of the rectangle vary with the distance from the reference axis. When

$$Z_B = -Z_T = -\frac{H}{2} = -6 \text{ in.},$$

we obtain

$$Z_A = 0,$$

$$Q = 0,$$

$$I = \frac{B}{3}\left[\left(\frac{H}{2}\right)^3 + \left(\frac{H}{2}\right)^3\right] = \frac{2}{24}BH^3 = \frac{BH^3}{12} = \frac{6}{12}(12)^3 = 864 \text{ in.}^4,$$

$$Z_Q = \infty.$$

We must, indeed, multiply the area moment $Q = 0$ by the infinite distance Z_Q to obtain a moment of inertia $I = QZ_Q$ other than zero. When

$$Z_B = 0 \text{ and } Z_T = H = 12 \text{ in.},$$

we obtain

$$Z_A = \frac{H}{2} = 6 \text{ in.},$$

$$Q = \frac{B}{2}Z_T^2 = \frac{BH^2}{2} = \frac{6}{2}(12)^2 = 432 \text{ in.}^3,$$

$$I = \frac{B}{3}Z_T^3 = \frac{B}{3}H^3 = \frac{6}{3}(12)^3 = 3456 \text{ in.}^4,$$

$$Z_Q = \frac{2}{3}Z_T = \frac{2H}{3} = 8 \text{ in.}$$

When

$$Z_B = 100 \text{ in.},$$

we obtain

$$Z_A = Z_B + \frac{H}{2} = 100 + 6 = 106 \text{ in.},$$

$$Z_Q = \frac{2}{3}\left[\frac{(112)^3 - (100)^3}{(112)^2 - (100)^2}\right] = 106 \text{ in.},$$

which confirms that Z_A and Z_Q can be used interchangeably when the depth of the area (here 6 in.) is small in comparison to its distance from the reference axis (here 106 in.).

As additional examples, the properties I and Z_Q are calculated for the triangular area $BH/2$ and the trapezoid in Figure 9.3*b*. The Cartesian y-axis is chosen as reference axis.

$$dQ_y = z\,dA = zb\,dz = z\left(\frac{B}{H}z\right)dz = \frac{B}{H}z^2\,dz,$$

$$dI_y = z\,dQ = \frac{B}{H}z^3\,dz.$$

The trapezoid therefore has these moments around the y-axis

$$Q_y = \int_h^H dQ_y = \frac{B}{H}\int_h^H z^2\,dz = \frac{B}{3H}(H^3 - h^3),$$

$$I_y = \int_h^H dI_y = \frac{B}{H}\int_h^H z^3\,dz = \frac{B}{4H}(H^4 - h^4).$$

The distance Z_Q from the reference axis y to the center of Q_y is

$$Z_Q = \frac{I_y}{Q_y} = \frac{3}{4}\left(\frac{H^4 - h^4}{H^3 - h^3}\right).$$

When $h = 0$, the trapezoid becomes the triangular area $BH/2$, for which

$$Q_y = \frac{BH^2}{3},$$

$$I_y = \frac{BH^3}{4},$$

$$Z_Q = \frac{3H}{4}.$$

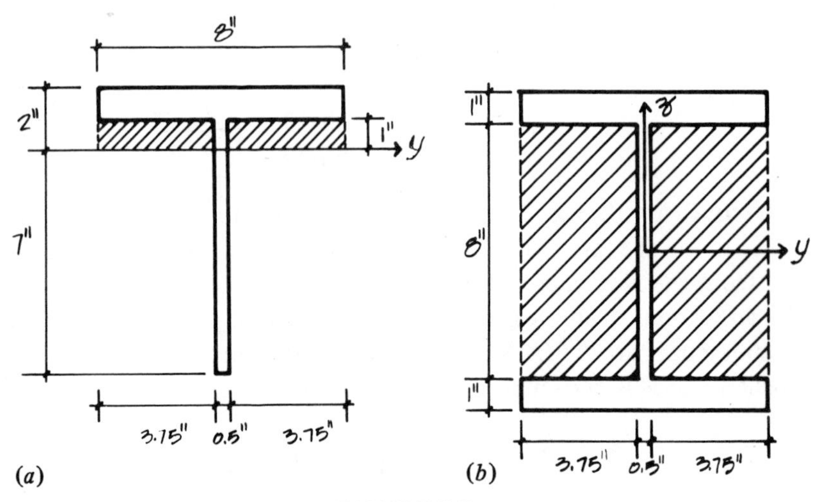

FIGURE 9.5

Properties already obtained as results of integrations can be used to calculate numerically the analogous properties for other cross sections. The T section in Figure 9.3a illustrates calculating the properties Q, I, and Z_Q. The horizontal centroidal axis has an elevation

$$z_b = 7 \text{ in.}$$

on the bottom fibers of the section, and it is used as the reference axis.

The T section is considered to be made up of three rectangles with their base on the centroidal axis y (Figure 9.5a). One rectangle is the web part below y. It has a 0.5-in. base on the y-axis and a 7-in. depth below the y-axis. A second rectangle has an 8-in. base on the y-axis and a 2-in. depth above the y-axis. The third rectangle is a combination of the two shaded cavities on the sides of the web. The 7.5-in. base of the rectangle is on the y-axis, and the 1-in. depth is above the y-axis. Applying the results obtained for the rectangle in Figure 9.4a where $Z_B = 0$, we have, for rectangle no. 1,

$$Q_{1y} = \frac{B_1 H_1^2}{2} = \frac{0.5}{2}(7)^2 = 12.25 \text{ in.}^3,$$

$$I_{1y} = \frac{B_1 H_1^3}{3} = \frac{0.5}{3}(7)^3 = 57.17 \text{ in.}^4,$$

$$Z_{Q1} = \frac{2}{3}H_1 = \frac{2}{3}(7) = 4.67 \text{ in.}$$

For rectangle no. 2, we have

$$Q_{2y} = \frac{B_2}{2}H_2^2 = \frac{8}{2}(2)^2 = 16 \text{ in.}^3,$$

$$I_{2y} = \frac{B_2}{3}(H_2)^3 = \frac{8}{3}(2)^3 = 21.33 \text{ in.}^4,$$

$$Z_{Q2} = \frac{2}{3}H_2 = \frac{2}{3}(2) = 1.33 \text{ in.}$$

For rectangle no. 3,

$$Q_{3y} = \frac{B_3}{2}H_3^2 = \frac{7.5}{2}(1.0)^2 = 3.75 \text{ in.}^3,$$

$$I_{3y} = \frac{B_3}{3}(H_3)^3 = \frac{7.5}{3}(1.0)^3 = 2.5 \text{ in.}^4,$$

$$Z_{Q3} = \frac{2}{3}H_3 = \frac{2}{3}(1.0) = 0.67 \text{ in.}$$

For area no. 2 minus area no. 3, the upper part of the T section, we have

$$Q_{2,3y} = Q_{2y} - Q_{3y} = 16 - 3.75 = 12.25 \text{ in.}^3$$

$$I_{2,3y} = I_{2y} - I_{3y} = 18.83 \text{ in.}^4$$

$$Z_{Q2,3} = \frac{I_{2,3y}}{Q_{2,3y}} = 1.54 \text{ in.}$$

$Q_{2,3y}$ has the same magnitude as Q_{1y}. Their algebraic sum is, indeed, the moment of the total area around a centroidal axis, and it must be zero.

The sum

$$I_{1y} + I_{2,3y} = I_y = 76 \text{ in.}^4$$

is the total moment of inertia of the T section about its centroidal axis y. I_y measures the bending strength of the T section. The sum

$$Z_{Q1} + Z_{Q2,3} = Z = 6.2 \text{ in.}$$

is the distance from the center of the lower-area moment to the center of the upper-area moment. Z measures the shear strength of the T section. These statements are substantiated in the derivation of bending and shear stresses.

The properties of a cross section and their calculations can be conveniently organized as shown in the following table, which contains results and calculations for the T section in Figure 9.5*a*.

Partial Area i	B_i (in.)	H_i (in.)	A_i (in.^{2})	z_i (in.)	Q_i (in.^{3})	I_i (in.^{4})	Z_i (in.)
1	0.5	7	3.5	-3.5	-12.25	57.17	-4.67
2	8	2	16	1	16	21.33	1.33
3	7.5	1	-7.5	0.5	-3.75	-2.5	0.67
Σ			12		0	76	

As an additional example of the use of the properties obtained by integrating, we calculate the moment of inertia about the y- and z-axes of the symmetric I section shown in Figure 9.5*b*. To calculate I_z, we consider the I section made up of three rectangles (a web and two flanges) with their centroidal axes coinciding with the centroidal z-axis of the I section. The results obtained for the rectangle in Figure 9.4*a*, in the case $Z_B = -Z_T = -H/2$ applied to the I beam, give

$$I_Z = \frac{B_1}{12} H_1^3 + \frac{B_2}{12} H_2^3 + \frac{B_3}{12} H_3^3$$

$$= \frac{1}{12}(8)^3 + \frac{8}{12}(0.5)^3 + \frac{1}{12}(8)^3 = 85.4 \text{ in.}^4$$

To calculate I_y, we consider the I section made up of two rectangles: an outer rectangle with $B = 8$ in. and $H = 10$ in. and a rectangular cavity combining the rectangles on both sides of the web. The width of the cavity is 7.5 in., the height is 8 in. The two partial areas have the same horizontal centroidal axis as the I section. The I sec-

$$dA = 2\pi r\, dr.$$

Each point of this area has a radial coordinate

$$r + f\, dr,$$

of which $f\, dr$, a fraction of the infinitesimal thickness dr, is negligible in comparison with the finite radius r. The entire area dA has, therefore, a polar distance r from the center of the circle. Then,

$$dI_p = r^2\, dA = 2\pi r^3\, dr.$$

Integrating, we obtain

$$I_p = \int_A dI_p = 2\pi \int_{r=0}^{r=R} r^3\, dr = \frac{\pi}{2} R^4.$$

Because of the area's polar symmetry

$$I_z = I_y;$$

then,

$$I_p = I_z + I_y = 2I_z = 2I_y,$$

from which

$$I_z = I_y = \frac{I_p}{2} = \frac{\pi}{4} R^4.$$

For example, the circular section in Figure 9.7*b* with radius $R = 2$ in. has second-order moments

$$I_p = \frac{\pi}{2}(2)^4 = 25.1 \text{ in.}^4,$$

$$I_z = I_y = \frac{I_p}{2} = 12.6 \text{ in.}^4.$$

9.4. THE RADIUS OF GYRATION

A cross-sectional property, one that indicates how effectively the structural material is distributed on the section of compression members in order to minimize their tendency to buckle, is called the *minimum radius of gyration*, and it is defined as follows:

$$\rho_{\min} = \left(\frac{I_{\min}}{A}\right)^{1/2},$$

where $I_{\min}$ is the smallest of the centroidal moments of inertia of the cross section and A is the cross-sectional area.

Since $I_{\min}$ is measured in in.^{4} and A is measured in in.^{2}, $r_{\min}$ must be measured in inches, as in a distance. If we multiply the second power of $\rho_{\min}$ by A, we obtain

$$A\,\rho_{\min}^2 = I_{\min}.$$

Since an infinitesimal area dA, a distance $\rho_{\min}$ from a reference axis, has a moment of inertia $\rho_{\min}^2\,dA$ around that axis, it is possible to give $\rho_{\min}$ a graphic interpretation. Indeed, $\rho_{\min}$ can be viewed as the distance from the centroidal axis to a point such that the cross-sectional area A, condensed there to infinitesimal dimensions, would have the same second-order moment that it actually has. For example, the minimum centroidal moment of inertia of a rectangle with base $B = 3$ in. and height $H = 6$ in. is

$$I_{\min} = \frac{H}{12}(B)^3 = \frac{6}{12}(3)^3 = 13.5 \text{ in.}^4.$$

If the cross-sectional area $BH = 18$ in.^{2} is condensed to a point distant by $B/(12)^{1/2}$ in. from its centroidal axis, the moment of inertia is, again,

$$18\left[\frac{3}{(12)^{1/2}}\right]^2 = 13.5 \text{ in.}^4$$

$B/(12)^{1/2}$ is, therefore, the minimum radius of gyration of the rectangular section BH, and it is obtained from

$$\rho_{\min} = \left(\frac{I_{\min}}{A}\right)^{1/2} = \left(\frac{HB^3/12}{BH}\right)^{1/2}.$$

As an additional example, we consider a hollow circular section with outer radius $R = 9$ in. and inner radius $r = 8$ in. In this case,

$$I_{\min} = I_{\max} = I = \frac{\pi}{4}(R^4 - r^4),$$

$$A = \pi(R^2 - r^2)$$

$$\rho_{\min} = \rho_{\max} = \rho = \left(\frac{I}{A}\right)^{1/2} = \left(\frac{(\pi/4)(R^4 - r^4)}{\pi(R^2 - r^2)}\right)^{1/2} = \frac{1}{2}(R^2 + r^2)^{1/2}$$

$$= \frac{1}{2}(81 + 64)^{1/2} = 6 \text{ in.}$$

The radius of gyration ρ can be used to find the center C_Q of an area moment.

To substantiate this statement and show the procedure that leads to C_Q by using ρ, we consider the rectangular area in Figure 9.4*a*. Values for B, H, and Z_A are chosen as follows:

$$B = 4 \text{ in.},$$

$$H = 10 \text{ in.},$$

$$Z_A = 20 \text{ in.}$$

The theorem of Huygens states

$$I_y = I_c + AZ_A^2 = A\rho^2 + AZ_A^2 = AZ_A Z_Q.$$

Thus,

$$\rho^2 + Z_A^2 = Z_A Z_Q.$$

Using the identity

$$Z_Q = Z_A + (Z_Q - Z_A),$$

the equation becomes

$$\rho^2 + Z_A^2 = Z_A[Z_A + (Z_Q - Z_A)].$$

Expanding this, we get

$$\rho^2 + Z_A^2 = Z_A^2 + Z_A(Z_Q - Z_A);$$

then,

$$\rho^2 = Z_A(Z_Q - Z_A),$$

from which

$$Z_Q - Z_A = \frac{\rho^2}{Z_A}.$$

In the case we are discussing,

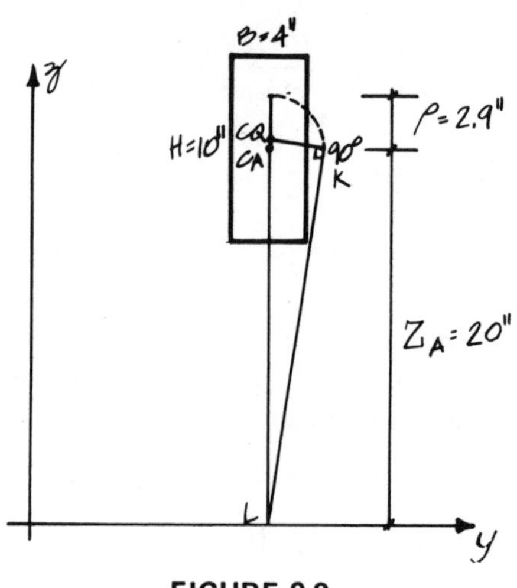

FIGURE 9.8

$$\rho^2 = \frac{H^2}{12} = \frac{10^2}{12} = 8.33 \text{ in.}^2.$$

Thus,

$$Z_Q - Z_A = \frac{8.33}{20} = 0.417 \text{ in.}$$

To obtain the same result with a graphic approach (Figure 9.8), the radius of gyration ρ perpendicular to reference axis y is drawn in scale. ρ is then flipped from its position on the centroidal axis perpendicular to y to a new position on the centroidal axis parallel to y. The segment KL is drawn from the end K of ρ to intersection L of reference axis y, with the centroidal axis perpendicular to y. A line perpendicular to KL drawn from K cuts the centroidal axis perpendicular to y at the center C_Q of the area moment. Indeed, triangle $C_Q KL$ has a 90° angle at K. Therefore, by the Euclidean theorem,

$$\overline{C_Q C_A}(\overline{C_A L}) = \overline{C_A K}^2;$$

since $\overline{C_A L}$ is Z_A, and $\overline{C_A K}$ is ρ,

$$\overline{C_Q C_A}(Z_A) = \rho^2,$$

which shows that $\overline{C_Q C_A}$ is $Z_Q - Z_A$ and it can be measured in scale.

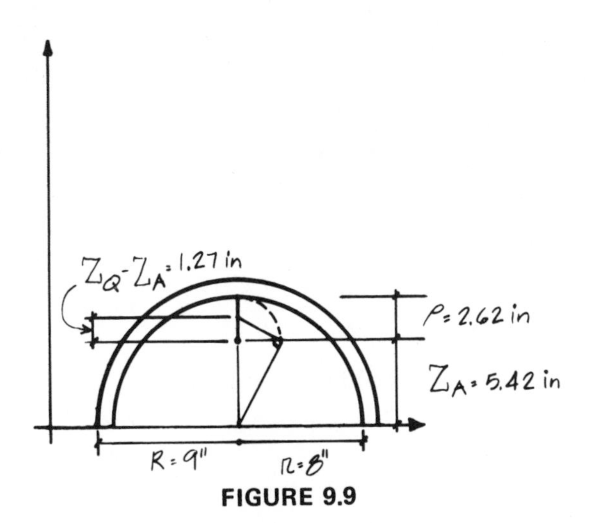

FIGURE 9.9

As a further example, we find the center of the moment of a semi-annular area around the diameter (Figure 9.9). The radii of the outer and inner circles are $R = 9$ in. and $r = 8$ in., respectively. The area of the half ring is

$$A = A_o - A_i = \frac{\pi}{2}(R^2 - r^2) = \frac{\pi}{2}(9^2 - 8^2) = 26.7 \text{ in.}^2$$

The distance from the center of A_o to the diameter is

$$\frac{4R}{3\pi} = \frac{4(9)}{3\pi} = 3.82 \text{ in.}$$

The analogous distance for A_i is

$$\frac{4r}{3\pi} = \frac{4(8)}{3\pi} = 3.4 \text{ in.}$$

Equating the moment of A around the diameter to the sum of the analogous moments of A_o and A_i, we obtain the distance from the center of A to the diameter

$$Z_A = \frac{(\pi/2)R^2(4R/3\pi) - (\pi/2)r^2(4r/3\pi)}{(\pi/2)(R^2 - r^2)} = (4/3\pi)\frac{(R^3 - r^3)}{(R^2 - r^2)} = 5.42 \text{ in.}$$

The centroidal moment of inertia is given by the Huygens theorem

$$I_c = I_y - AZ_A^2 = \frac{\pi}{8}(R^4 - r^4) - \frac{\pi}{2}(R^2 - r^2)Z_A^2$$

$$= 68 - 26.7(5.42)^2 = 183.5 \text{ in.}^4.$$

Then,

$$\rho = \left(\frac{Ic}{A}\right)^{1/2} = \left(\frac{183.5}{26.7}\right)^{1/2} = 2.62 \text{ in.}$$

Finally,

$$Z_Q - Z_A = \frac{\rho^2}{Z_A} = \frac{(2.62)^2}{5.42} = 1.27 \text{ in.}$$

In Figure 9.9, this result is obtained graphically.

PROBLEMS

9.1. Calculate the resultants of the parabolic loads on the beam in Figure 9.10*a*.

Solution. The resultant of the load on the half span is the area of parabola A_1 in Figure 9.10*c*

$$A_1 = \int_0^{l/2} g(x)\, dx$$

The expression for the parabolic load $g(x)$ is derived in Problem 5.2

$$g(x) = \left(\frac{4}{l^2} g_{\max}\right) x^2 - \left(\frac{4}{l} g_{\max}\right) x + g_{\max}.$$

Thus,

$$A_1 = \frac{4}{l^2} g_{\max} \int_0^{l/2} x^2\, dx - \frac{4}{l} g_{\max} \int_0^{l/2} x\, dx + g_{\max} \int_0^{l/2} dx.$$

Integrating, we obtain

$$A_1 = \frac{4}{3l^2} g_{\max} \frac{l^3}{8} - \frac{4}{2l} g_{\max} \frac{l^2}{4} + g_{\max} \frac{l}{2}$$

$$= \frac{l}{6} g_{\max} - \frac{l}{2} g_{\max} + \frac{l}{2} g_{\max} = \frac{1}{3}\frac{l}{2} g_{\max}.$$

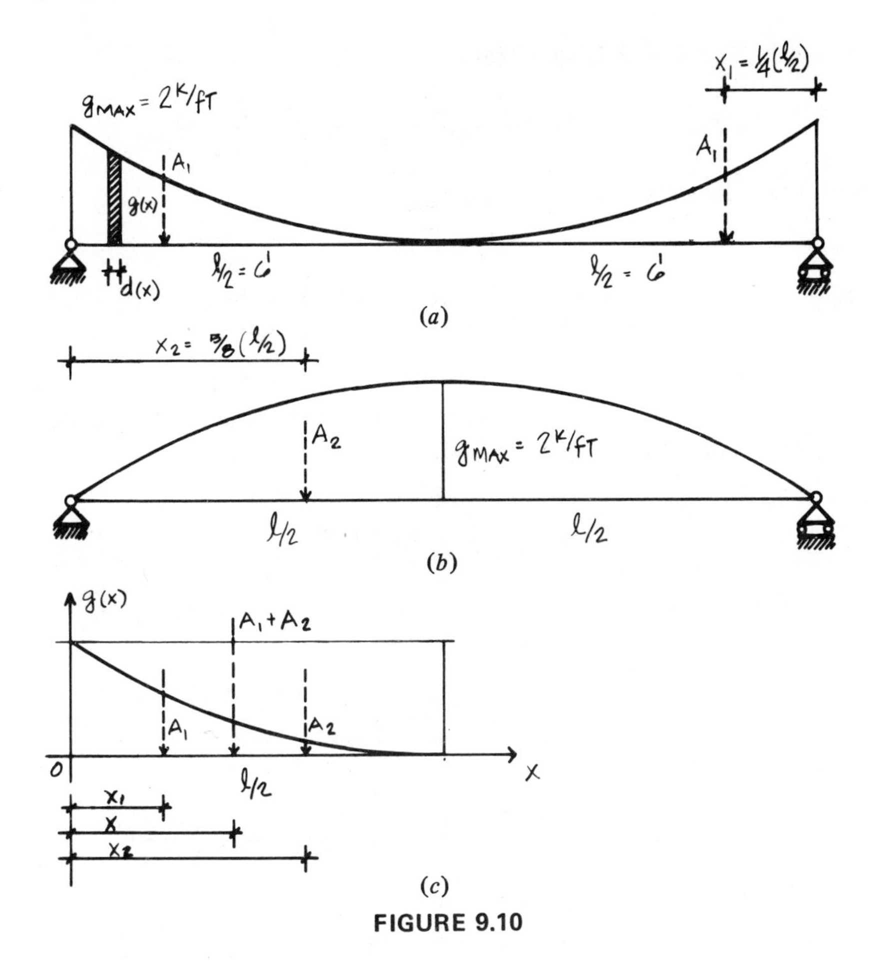

FIGURE 9.10

With the values specified for l and g_{max},

$$A_1 = 4 \text{ k}.$$

9.2. Calculate the resultants of parabolic loads on the beam in Figure 9.10*b*.

Solution. The resultant of the load on the half span is area A_2 in Figure 9.10*c*. Since the rectangular area of base $l/2$ and height g_{max} measures

$$A_1 + A_2 = \frac{l}{2} g_{max},$$

then from Problem 9.1, we obtain

$$A_2 = \frac{l}{2} g_{\max} - A_1 = \frac{l}{2} g_{\max} - \frac{1}{3}\frac{l}{2} g_{\max} = \frac{2}{3}\frac{l}{2} g_{\max}.$$

With the values specified for l and $g_{\max}$,

$$A_2 = 8 \text{ k}.$$

9.3. Calculate the distance X_1 from the end of the beam in Figure 9.10*a* to resultant A_1 of the load on the half span.

Solution. The moment of resultant A_1 around the beam end is equal to the sum of the moments of the elemental loads [shaded areas $g(x)\,dx$] around the same end.

$$A_1 X_1 = \int_0^{l/2} x g(x)\, dx$$

Replacing $g(x)$ with its expression (from Problem 9.1), we obtain

$$A_1 X_1 = \frac{4}{l^2} g_{\max} \int_0^{l/2} x^3\, dx - \frac{4}{l} g_{\max} \int_0^{l/2} x^2\, dx + g_{\max} \int_0^{l/2} x\, dx.$$

Integrating yields

$$A_1 X_1 = \frac{4}{\rho^2} g_{\max} \left(\frac{1}{4}\right) \frac{l^4}{16} - \frac{4}{l} g_{\max} \left(\frac{1}{3}\right) \frac{l^3}{8} + g_{\max} \left(\frac{1}{2}\right) \frac{l^2}{4}$$

$$= \frac{l^2}{16} g_{\max} \left(1 - \frac{8}{3} + 2\right) = \frac{l^2}{48} g_{\max}.$$

Replacing A_1 with its expression from Problem 9.1 gives

$$X_1 \frac{l}{6} g_{\max} = \frac{l^2}{48} g_{\max}.$$

Thus,

$$X_1 = \frac{1}{4}\left(\frac{l}{2}\right).$$

9.4. Calculate the distance X_2 from the end of the beam in Figure 9.10*b* to resultant A_2 of the load on the half span.

Solution. Operating for convenience on Figure 9.10*c*, we equate the moment of the rectangular load (area $A_1 + A_2$) around the origin 0 of the coordinates to the sum of the moments of A_1 and A_2 around the origin

$$(A_1 + A_2)\, X = A_1 X_1 + A_2 X_2.$$

Replacing A_1, A_2, X, and X_1 with their expressions from previous problems, we obtain

$$\frac{l}{2} g_{\max} \left(\frac{l}{2}\right) \frac{1}{2} = \frac{1}{3}\left(\frac{l}{2}\right) g_{\max} \left(\frac{l}{2}\right) \frac{1}{4} + \frac{2}{3}\left(\frac{l}{2}\right) g_{\max}\, (X_2),$$

from which, after dividing by $(l/2)\, g_{\max}$,

$$X_2 = \frac{3}{2}\left(\frac{1}{2} - \frac{1}{12}\right) \frac{l}{2} = \frac{5}{8}\left(\frac{l}{2}\right).$$

9.5. A precast, prestressed concrete beam has the cross section shown in Figure 9.11 by solid lines. Find the horizontal centroidal axis of the beam section.

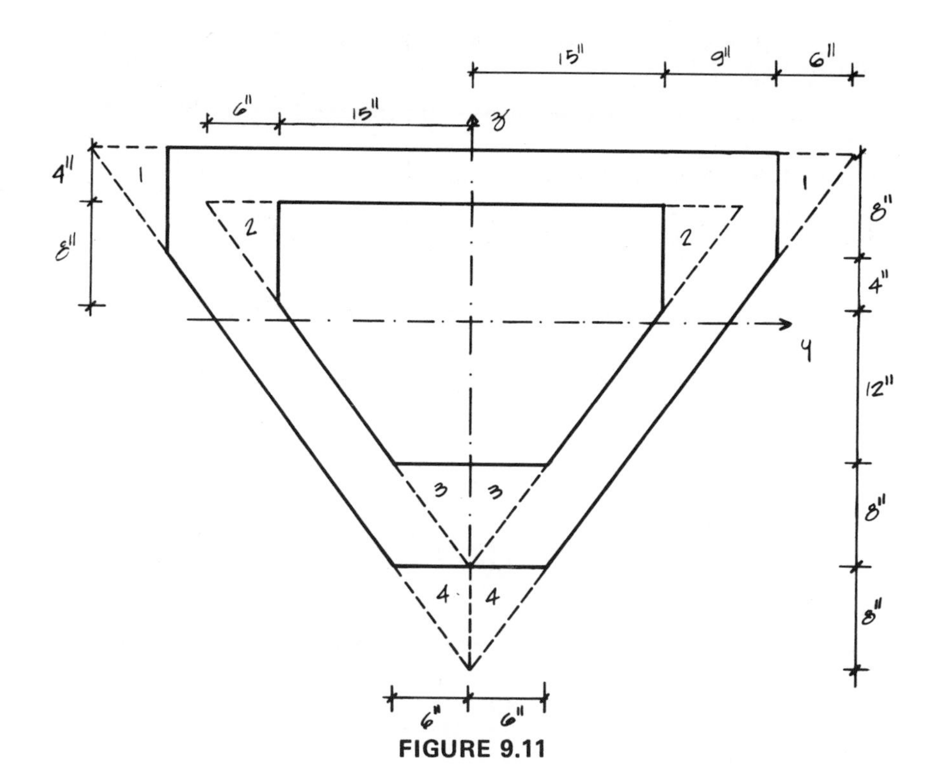

FIGURE 9.11

Solution. The sectional area is made of six triangles: the outer triangle with dotted corners, with area

$$A_6 = 0.5(60)\,40 = 1200 \text{ in.}^2.$$

The inner triangular cavity with dotted corners, with area

$$A_5 = -0.5(42)\,28 = -588 \text{ in.}^2.$$

Cavities 1 and 4 having areas, respectively, of

$$A_1 = -0.5(12)\,8 - -48 \text{ in.}^2,$$

$$A_4 = -48 \text{ in.}^2.$$

Fillings 2 and 3 having areas, respectively, of

$$A_2 = 48 \text{ in.}^2,$$

$$A_3 = 48 \text{ in.}^2.$$

Total area A is given by

$$A = \sum_{i=1}^{6} A_i = A_1 + A_2 + A_3 + A_4 + A_5 + A_6$$

$$= -48 + 48 + 48 - 48 - 588 + 1200 = 612 \text{ in.}^2.$$

The moment of A_6 around the top fibers of the section is

$$A_6 z_{6T} = \frac{1200}{3}(40) = 16{,}000 \text{ in.}^3.$$

The moments of the other areas are calculated similarly:

$$A_5 z_{5T} = -588\left(4 + \frac{28}{3}\right) = -7840 \text{ in.}^3,$$

$$A_4 z_{4T} = -48\left(32 + \frac{8}{3}\right) = -1664 \text{ in.}^3,$$

$$A_3 z_{3T} = 48\left(24 + \frac{8}{3}\right) = 1280 \text{ in.}^3,$$

$$A_2 z_{2T} = 48\left(4 + \frac{8}{3}\right) = 320 \text{ in.}^3,$$

$$A_1 z_{1T} = \frac{-48}{3}(8) = -128 \text{ in.}^3.$$

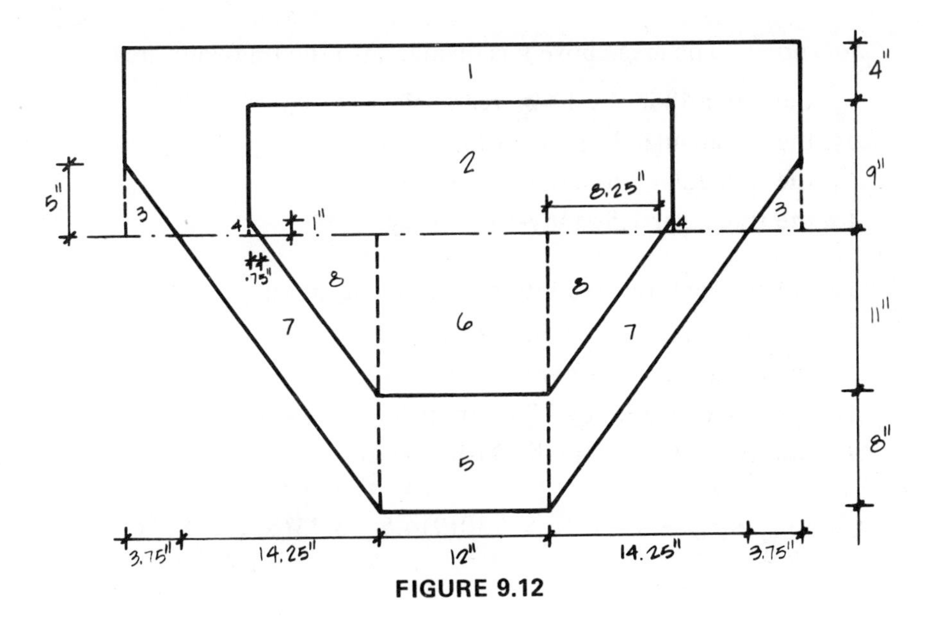

FIGURE 9.12

Thus,

$$\sum_{i=1}^{6} A_i z_{iT} = -128 + 320 + 1280 - 1664 - 7840 + 16{,}000 = 7968 \text{ in.}^3.$$

The distance from the top fibers to the center of the section is

$$z_{\text{top}} = \frac{\sum_{i=1}^{6} A_i z_{iT}}{\sum_{i=1}^{6} A_i} = \frac{7968}{612} = 13 \text{ in.}$$

9.6. For the cross section in Problem 9.5, (Figure 9.12) find:

1. The static moment Q_A around y of the area above y;
2. The static moment Q_B around y of the area below y;
3. The moment of inertia I_A around y of the area above y;
4. The moment of inertia I_B around y of the area below y;
5. The total moment of inertia I of the section around y;
6. The distance Z_A from y to the center of Q_A;
7. The distance Z_B from y to the center of Q_B;
8. The distance Z from the center of Q_A to the center of Q_B.

Solution. The area above y is divided into the following areas:

A_1, the outer 48-in.-by-13-in. rectangle.
A_2, the rectangular 30-in.-by-9-in. cavity.
A_3, the triangular 7.5-in.-by-5-in. cavity.
A_4, the triangular 1.5-in.-by-1-in. filling.

The area below y is divided into the following areas:

A_5, the outer 12-in-by-19-in. rectangle.
A_6, the rectangular 12-in.-by-11-in. cavity.
A_7, the outer 28.5-in.-by-19-in. triangle.
A_8, the triangular 16.5-in.-by-11-in. cavity.

$$Q_A = \sum_{i=1}^{4} A_i z_{iA} = 48(13)6.5 - 30(9)4.5 - 3.75(5)\tfrac{5}{3} + 0.75(1)\tfrac{1}{3}$$

$$= 2810 \text{ in.}^3,$$

$$Q_B = \sum_{i=5}^{8} A_i z_{iB} = 12(19)9.5 - 12(11)5.5 + 14.25(19)\tfrac{19}{3}$$

$$- 8.25(11)\tfrac{11}{3} = 2822 \text{ in.}^3.$$

The discrepancy between Q_A and Q_B is 0.43%, a negligible figure.

$$I_A = \sum_{i=1}^{4} I_i = \frac{48}{3}(13)^3 - \frac{30}{3}(9)^3 - \frac{7.5}{12}(5)^3 + \frac{1.5}{12}(1)^3 = 27{,}784 \text{ in.}^4,$$

$$I_B = \sum_{i=5}^{8} I_i = \frac{12}{3}(19)^3 - \frac{12}{3}(11)^3 + \frac{28.5}{12}(19)^3 - \frac{16.5}{12}(11)^3$$

$$= 36{,}572 \text{ in.}^4,$$

$$I = I_A + I_B = 64{,}536 \text{ in.}^4,$$

$$Z_A = \frac{I_A}{Q_A} = \frac{27{,}784}{2810} = 9.9 \text{ in.},$$

$$Z_B = \frac{I_B}{Q_B} = \frac{36{,}572}{2822} = 13 \text{ in.},$$

$$Z = Z_A + Z_B = 22.9 \text{ in.}$$

9.7. Find the following properties for the cross section in Figure 9.11:

Moment Q_z of the half section around the z-axis.
Inertia moment I_z around the z-axis.
Distance Y from the center of the left moment Q_z to the center of the right moment Q_z.

Solution. The outer triangle with a 40-in. base on the z-axis and 30-in. height parallel to y has the moment around z (o for outer)

$$Q_{oz} = \frac{1}{2}(40)(30)(10) = 6000 \text{ in.}^3.$$

The inner cavity with a 28-in. base on the z-axis and 21-in. height parallel to y has the moment around z (i for inner)

$$Q_{iz} = -\frac{1}{2}(28)(21)(7) = -2058 \text{ in.}^3.$$

The moment of the cavity 1 is

$$Q_{1z} = -\frac{1}{2}(8)(6)(24+2) = 624 \text{ in.}^3.$$

The moment of filling 2 is

$$Q_{2z} = \frac{1}{2}(8)(6)(15+2) = 408 \text{ in.}^3.$$

Filling 3 replaces cavity 4. The moment of half section around the z-axis is, then,

$$Q_z = Q_{oz} - Q_{iz} - Q_{1z} + Q_{2z} = 6000 - 2058 - 624 + 408 = 3726 \text{ in.}^3.$$

The combined moment of inertia around z of the two triangles with a 40-in. base on the z-axis and a 30-in. height parallel to y is

$$\frac{40}{12}(30)^3 2 = 180{,}000 \text{ in.}^4.$$

The combined moment of inertia around z of the triangular cavities 1 with an 8-in. base parallel to z and a 6-in. altitude parallel to y is

$$-2\left[\frac{8}{36}(6)^3 + 24(26)^2\right] = -32{,}544 \text{ in.}^4$$

The moment of inertia around z of the rectangular cavity with an 8-in. base parallel to z and a 30-in. height parallel to y is

$$-\frac{8}{12}(30)^3 = -18{,}000 \text{ in.}^4.$$

Filling 3 and cavity 4 can exchange places without changing the value of I_z. The combined moment of inertia around z of the two triangular cavities with a 20-in. base on the z-axis and a 15-in. height parallel to y is

$$-(2)\left(\frac{20}{12}\right)(15)^3 = -11{,}250 \text{ in.}^4.$$

The total moment of inertia I_z is, then,

$$I_z = 180{,}000 - 32{,}544 - 18{,}000 - 11{,}250 = 118{,}206 \text{ in.}^4.$$

Finally, distance Y is

$$Y = \frac{I_z}{Q_z} = \frac{118{,}206}{3726} = 31.7 \text{ in.}$$

9.8. Find the moment of inertia I_y of the cross section in Figures 9.11 and 9.12 around the y-axis using the Huygens theorem.

Solution. The outer triangle has a moment of inertia I_{y_o} around its own centroidal axis y_o parallel to y (self-moment of inertia)

$$I_{y_o} = \frac{60}{36}(40)^3 = 106{,}667 \text{ in.}^4.$$

The center of the outer triangle is 13.33 in. below the top fibers of the cross section. The y-axis is 13 in. below the same fibers. Thus, the distance from y_o to y is 0.33 in. The outer triangle has a moment of inertia I_o around the y-axis

$$I_o = 106{,}667 + \frac{1}{2}(60)(40)(0.33)^2 = 106{,}780 \text{ in.}^4.$$

The inner triangular cavity with dotted corners has a moment of inertia I_{y_i} about its own centroidal axis y_i parallel to y

$$I_{y_i} = \frac{42}{36}(28)^3 = 25{,}610.7 \text{ in.}^4.$$

The center of the cavity is also 13.33 in. below the top fibers; hence, the distance from y_i to y is also 0.33 in. The moment of inertia I_i of the triangular cavity around y is

$$I_i = 25{,}610.7 + \frac{1}{2}(42)(28)(0.33)^2 = 25{,}676 \text{ in.}^4.$$

The self-moment of inertia of each of the corners 1, 2, 3, and 4 is

$$\frac{6}{36}(8)^3 = 85.3 \text{ in.}^4.$$

The distance from the centroidal axis y_1 of the triangles labeled 1 to the y-axis is

$$13 - \frac{8}{3} = 10.33 \text{ in.}$$

Thus, the moment of inertia I_1 of these triangles around y is

$$I_1 = 2(85.3) + (2)(24)(10.33)^2 = 5296 \text{ in.}^4.$$

Similarly, the distance from y_2 to y is

$$13 - \left(\frac{8}{3} + 4\right) = 6.33 \text{ in.},$$

and the moment of inertia I_2 of the triangles labeled 2 around y is

$$I_2 = 2(85.3) + (2)(24)(6.33)^2 = 2096 \text{ in.}^4.$$

The distance from y_3 to y is

$$19 - \frac{2}{3}(8) = 13.67 \text{ in.}$$

Therefore,

$$I_3 = 2(85.3) + (2)(24)(13.67)^2 = 9136 \text{ in.}^4.$$

The distance from y_4 to y is

$$19 + \frac{8}{3} = 21.67 \text{ in.}$$

Thus,

$$I_4 = 2(85.3) + (2)(24)(21.67)^2 = 22{,}704 \text{ in.}^4.$$

The total moment of inertia I_y is

$$I_y = I_o - I_i - I_1 + I_2 + I_3 - I_4$$
$$= 106{,}780 - 25{,}676 - 5296 + 2096 + 9136 - 22{,}704$$
$$= 64{,}336 \text{ in.}^4.$$

The discrepancy in the answer to Problem 9.6 is 0.03%–negligible.

9.9. Find the moment of inertia I_z of the cross section in Figures 9.11 and 9.12 around the z-axis using the Huygens theorem.

Solution for one half section. The outer triangle with a 40-in. base on the z-axis and a 30-in. height parallel to y has a moment of inertia around its centroidal axis z_o parallel to z (self-moment of inertia).

$$I_{z_o} = \frac{40}{36}(30)^3 = 30{,}000 \text{ in.}^4.$$

The distance from z_o to z is 10 in. Therefore, the moment of inertia I_o of the outer triangle around z is

$$I_o = 30{,}000 + 600(10)^2 = 90{,}000 \text{ in.}^4.$$

Interchanging filling 3 and cavity 4 does not alter I_z.

The self-moment of inertia I_{z_i} of the inner triangular cavity with a 28-in. base on the z-axis and a 21-in. height parallel to y is

$$I_{z_i} = \frac{28}{36}(21)^3 = 7203 \text{ in.}^4.$$

The distance from z_i to z is 7 in.; thus,

$$I_i = 7203 + 294(7)^2 = 21609 \text{ in.}^4.$$

The self-moment of inertia of each of the corners 1 and 2 is

$$I_{z_1} = I_{z_2} = \frac{8}{36}(6)^3 = 48 \text{ in.}^4.$$

The distance from z_1 to z is 26 in.; thus,

$$I_1 = 48 + 24(26)^2 = 16{,}272.$$

The distance from z_2 to z is 17 in.; thus,

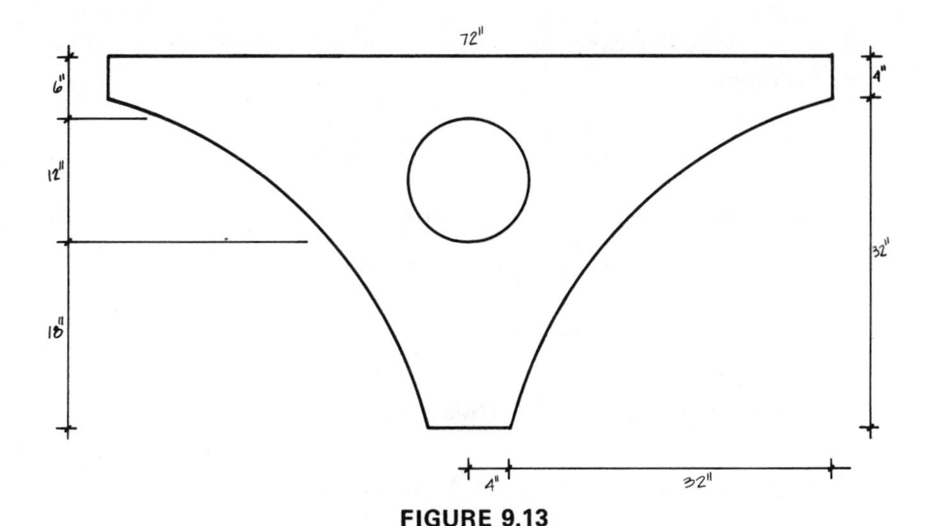

FIGURE 9.13

$$I_2 = 48 + 24(17)^2 = 6984 \text{ in.}^4.$$

The moment of inertia of the half section around z is

$$\frac{1}{2} I_z = I_o - I_i - I_1 + I_2 = 90{,}000 - 21{,}609 - 16{,}272 + 6984$$

$$= 59103 \text{ in.}^4,$$

from which

$$I_z = 118{,}206 \text{ in.}^4,$$

a result identical to that in Problem 9.7.

9.10. The precast pretensioned beams of a vaulted ceiling have the cross section shown in Figure 9.13. Find the horizontal centroidal axis and the moment of inertia of the cross section around this axis.

Solution. The area of the cross section is the algebraic sum of the following partial areas:

A_1, an outer rectangle with a 72-in. horizontal base and a 36-in. height

$$A_1 = 72(36) = 2592 \text{ in.}^2.$$

A_2, a semicircular cavity combination of the two quadrants with a 32-in. radius

$$A_2 = \frac{\pi}{2}(32)^2 = 1608.5 \text{ in.}^2.$$

A_3, a circular cavity with a 6-in. radius

$$A_3 = \pi(6)^2 = 113 \text{ in.}^2.$$

Thus,

$$A = \sum_{i=1}^{3} A_i = 2592 - 1608.5 - 113 = 870.5 \text{ in.}^2.$$

The distance from the center of A_1 to the bottom fibers is

$$z_{1b} = 18 \text{ in.}$$

With similar notations,

$$z_{2b} = \frac{4(32)}{3\pi} = 13.58 \text{ in.},$$

$$z_{3b} = 24 \text{ in.}$$

The static moment of A_1 around the bottom fibers is

$$A_1 z_{1b} = 2592(18) = 46{,}656 \text{ in.}^3.$$

Using similar notations,

$$A_2 z_{2b} = 1608.5(13.58) = 21{,}843.4 \text{ in.}^3,$$

$$A_3 z_{3b} = 113(24) = 2712 \text{ in.}^3.$$

The distance from the centroid of the section to the bottom fibers is, thus,

$$z_b = \frac{\sum_{i=1}^{3} A_i z_{ib}}{\sum_{i=1}^{3} A_i} = \frac{46{,}656 - 21{,}843.4 - 2712}{870.5} = 25.4 \text{ in.}$$

The moment of inertia I_b of the cross section around the bottom fibers is the algebraic sum of moments I_{1b}, I_{2b}, and I_{3b} of partial areas A_1, A_2, and A_3, respectively.

$$I_{1b} = \frac{72}{3}(36)^3 = 1{,}119{,}744 \text{ in.}^4,$$

$$I_{2b} = \frac{1}{2}\frac{\pi}{4}(32)^4 = 411{,}775 \text{ in.}^4.$$

By the Huygens theorem,

$$I_{3b} = \frac{\pi}{4}(6)^4 + \pi(6)^2(24)^2 = 1018 + 65{,}144 = 66{,}162 \text{ in.}^4.$$

Thus,

$$I_b = 1{,}119{,}744 - 411{,}775 - 66{,}162 = 641{,}807 \text{ in.}^4.$$

By the Huygens theorem, the moment of inertia I_b also equals the sum of the centroidal moment of inertia I_y with the transfer term Az_b^2

$$I_b = I_y + Az_b^2.$$

Replacing symbols with numbers,

$$641{,}807 = I_y + 870.5(25.4)^2,$$

from which

$$I_y = 641{,}807 - 870.5(25.4)^2 = 80{,}195 \text{ in.}^4.$$

9.11. Find the moment of inertia I_z around the axis of symmetry for the cross section in Problem 9.10.

Solution. The section is considered to comprise a 36-in.-by-72-in. outer rectangle A_1, a circular cavity A_3 with a 6-in. radius, and two hollow quadrants with a 32-in. radius, each labeled A_2. The partial moments of inertia are

$$I_1 = \frac{36}{12}(72)^3 = 1{,}119{,}744 \text{ in.}^4,$$

$$I_3 = \frac{\pi}{4}(6)^4 = 1018 \text{ in.}^4.$$

The self-moment of inertia I_{z_2} of a quadrant around its vertical centroidal axis z_2 is obtained by using the Huygens theorem, the moment of inertia around the vertical diameter, and the distance $4(32)/3\pi$ from the vertical diameter to the centroid of the quadrant

$$I_{z_2} = \frac{1}{4}\left(\frac{\pi}{4}\right)(32)^4 - \frac{1}{4}\pi(32)^2\left[\frac{4(32)}{3\pi}\right]^2 = 205{,}887 - 148{,}343$$

$$= 57{,}544 \text{ in.}^4.$$

The partial moment of inertia I_2 is now obtained by using the Huygens theorem, I_{z_2}, and the distance y_2 from z_2 to z.

$$y_2 = 36 - \frac{4(32)}{3\pi} = 36 - 13.58 = 22.42 \text{ in.}$$

Thus,

$$\frac{1}{2}I_2 = I_{z_2} + \frac{1}{4}\pi(32)^2 y_2^2 = 57{,}544 + 404{,}260 = 461{,}804 \text{ in.}^4,$$

from which

$$I_2 = 923{,}608 \text{ in.}^4.$$

Finally,

$$I_z = I_1 - I_2 - I_3 = 1{,}119{,}744 - 923{,}608 - 1018 = 195{,}118 \text{ in.}^4.$$

9.12. Find the polar moment of inertia around the centroid for the cross section in Figure 9.11.

Solution. From Problem 9.6,

$$I_y = 64{,}356 \text{ in.}^4,$$

from Problem 9.7,

$$I_z = 118{,}206;$$

then,

$$I_p = I_y + I_z = 64{,}356 + 118{,}206 = 182{,}562 \text{ in.}^4.$$

9.13. Find the polar moment of inertia around the centroid for the cross section in Figure 9.13.

Solution. From Problem 9.10,

$$I_y = 80{,}195 \text{ in.}^4;$$

from Problem 9.11,

$$I_z = 195{,}118 \text{ in.}^4;$$

thus,

$$I_p = I_y + I_z = 80{,}195 + 19{,}518 = 275{,}313 \text{ in.}^4.$$

9.14. Find the least radius of gyration for the cross section in Figure 9.11.

Solution. From Problems 9.5, 9.6, and 9.7,

$$\begin{aligned} A &= 612 \text{ in.}^2, \\ I_y &= 64{,}356 \text{ in.}^4, \\ I_z &= 118{,}206 \text{ in.}^4. \end{aligned}$$

The radius of gyration ρ_z colinear with the z-axis is

$$\rho_z = \left(\frac{I_y}{A}\right)^{1/2} = \left(\frac{64{,}356}{612}\right)^{1/2} = 10.25 \text{ in.}$$

The radius of gyration ρ_y colinear with the y-axis is

$$\rho_y = \left(\frac{I_y}{A}\right)^{1/2} = \left(\frac{18{,}206}{612}\right)^{1/2} = 13.9 \text{ in.}$$

Thus, ρ_z is the least radius of gyration.

9.15. Find the least radius of gyration for the cross section in Figure 9.13.

Solution. From Problems 9.10 and 9.11,

$$\begin{aligned} A &= 870.5 \text{ in.}^2, \\ I_y &= 80{,}195 \text{ in.}^4, \\ I_z &= 195{,}118 \text{ in.}^4; \end{aligned}$$

thus,

$$\rho_z = \left(\frac{I_y}{A}\right)^{1/2} = \left(\frac{80{,}195}{870.5}\right)^{1/2} = 9.6 \text{ in.}$$

$$\rho_y = \left(\frac{I_z}{A}\right)^{1/2} = \left(\frac{195{,}118}{870.5}\right)^{1/2} = 14.97 \text{ in.}$$

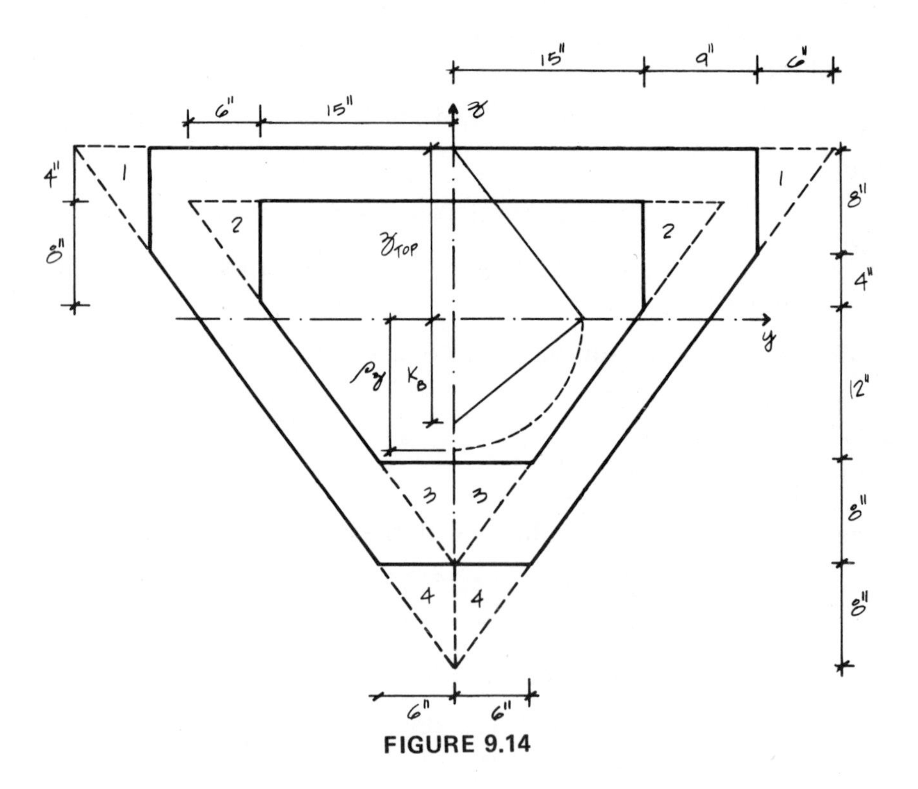

FIGURE 9.14

The radius of gyration ρ_z colinear with the z-axis is the least radius of gyration.

9.16. For the cross section in Figure 9.11, find the center of its area moment around the top line. (This point belongs to the Kern region, another geometric property of the cross section.)

Solution (see Figure 9.14). From Problems 9.14 and 9.5,

$$\rho_z = 10.25 \text{ in.},$$
$$z_{\text{top}} = 13 \text{ in.}$$

Then,

$$k_B = \frac{\rho_z^2}{z_{\text{top}}} = \frac{(10.25)^2}{13} = 8.08 \text{ in.}$$

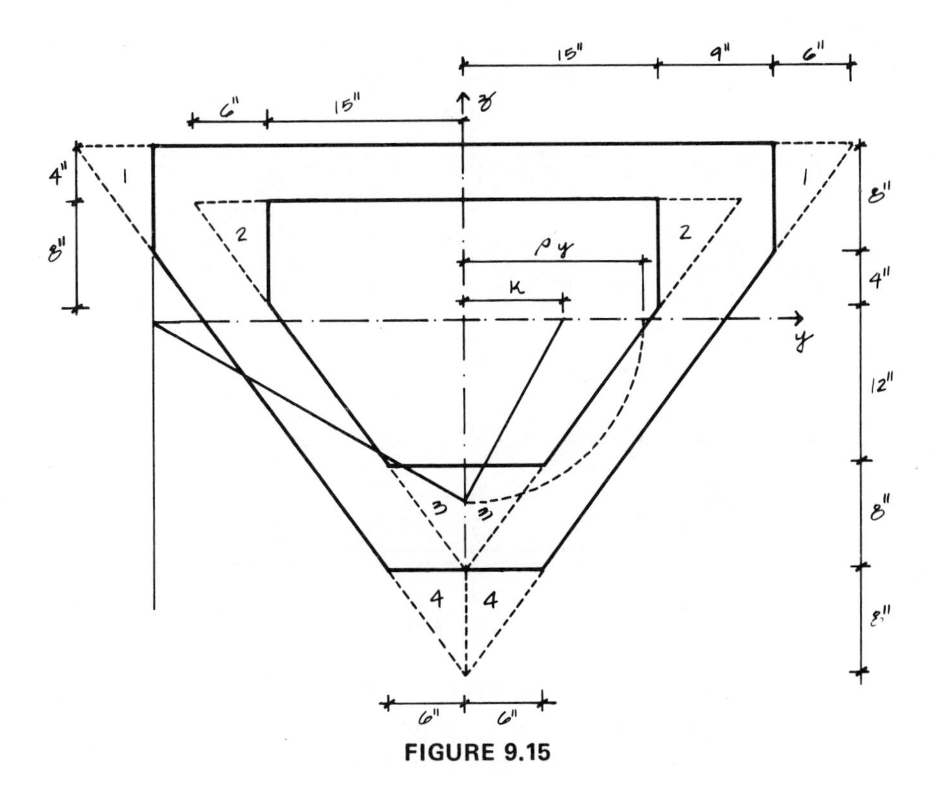

FIGURE 9.15

9.17. Find the center of the area moment around the vertical side (a point of the section's Kern region) for the cross section in Figure 9.11.

Solution (see Figure 9.15). The horizontal coordinate of the vertical side measures 24 in. From Problem 9.14,

$$\rho_y = 13.9 \text{ in.};$$

thus,

$$k = \frac{(13.9)^2}{24} = 8.05 \text{ in.}$$

9.18. For the cross section in Figure 9.16 find the following properties:

1. The moment Q_y of the area of a half section around the

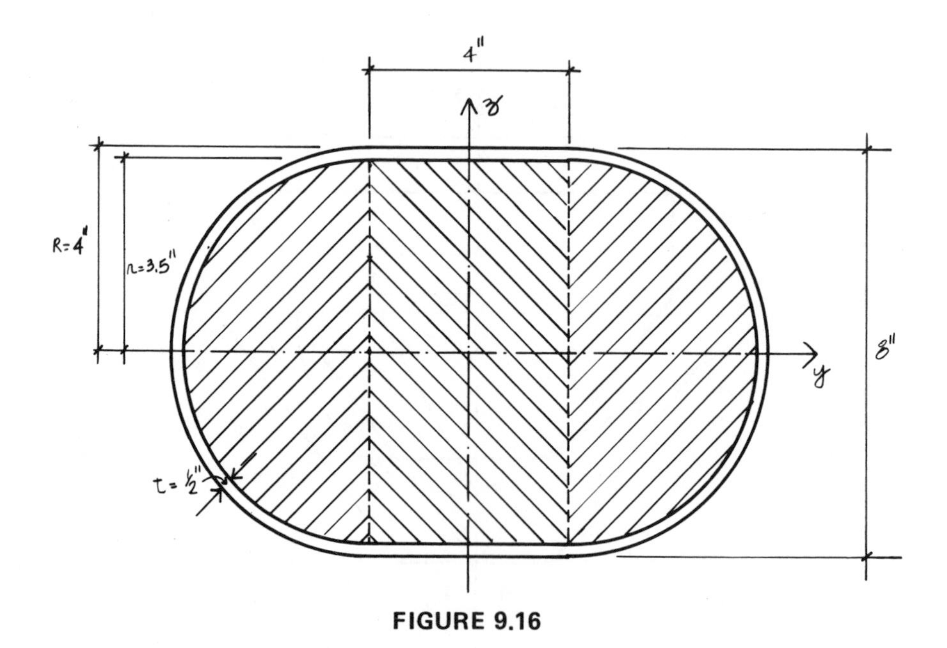

FIGURE 9.16

y-axis. (Hint: The distance from the centroid of a circular quadrant to the diameter is $4r/3\pi$).

2. The moment of inertia of one half section around the y-axis.
3. The distance from the center of Q_y to the y-axis.

Solution.

1. $Q_y = 21.6$ in.^{3},
2. $I_y/2 = 69.8$ in.^{4},
3. $Z/2 = 3.23$ in.

9.19. Verify the formulas given by the *AITC Timber Construction Manual* on page 7–32 for the properties of four different cross sections following these steps:

1. By giving numerical values to external symbols.
2. By obtaining numerical values of properties with the manual's formulas and your numerical values.
3. By independently obtaining numerical values of the same properties.

INDEX